Through the Earth

The Engineering of Nikola Tesla

1856–1943

For inquiries: info@robertwharrison.com.au

ISBN: 978-0-6486882-7-3

This is a work of nonfiction. The narrative is based on published primary sources, patents, and the author's interpretation of the historical and scientific record. Where the author offers analysis or speculation, this is clearly indicated in the text.

THROUGH THE EARTH

THE ENGINEERING OF NIKOLA TESLA

Robert W. Harrison

"Invariably my device works as I conceived that it should, and the experiment comes out exactly as I planned it. In twenty years there has not been a single exception."

— Nikola Tesla, My Inventions, 1919

CONTENTS

PROLOGUE

There are two Nikola Teslas in the popular imagination, and neither of them is real.

The first is the martyred genius: a man of supernatural brilliance who could have given the world free energy, defeated only by the greed of powerful men who couldn't allow it. In this version, Tesla is a secular saint, too pure for the world he was born into, crushed by forces that feared what he might unleash. It's a compelling story. It has driven a remarkable cultural revival over the past two decades, turned Tesla into one of the most recognizable names in the history of science, and inspired an electric car company, a unit of measurement, and an inexhaustible supply of internet memes. It is also, in important ways, wrong. Tesla was not crushed by villains. He was an extraordinarily poor manager of money and people, and his greatest project failed in large part because of decisions he made himself.

The second Tesla is the cautionary tale: the eccentric who started strong, overreached, and descended into delusion — claiming death rays, talking to pigeons, and dying alone in a hotel room. In this version, Tesla is a warning about what happens when brilliance goes unchecked by practicality. It's the version preferred by people who find the first version embarrassing, and it has the advantage of being easy to tell. It is also, in more important ways, wrong. Tesla's later years were not a descent into madness. His engineering principles remained sound. His mind remained sharp. What failed was not his intellect but his circumstances.

This book is an attempt to set both versions aside and look at what actually happened.

* * *

The strange thing about Nikola Tesla is that almost nobody takes him seriously as an engineer. His admirers tend to treat him as a mystic, a prophet, a figure of almost religious significance. His detractors tend to treat him as a showman who got lucky with alternating current and spent the rest of his career chasing fantasies. Both camps skip over the part that matters most: the work.

Tesla's engineering record is, by any honest measure, extraordinary. He designed the rotating magnetic field, the polyphase alternating current system, the Tesla coil, the magnifying transmitter, the bladeless turbine, the Tesla valve, and the first remote-controlled vehicle, among dozens of other inventions. He held hundreds of patents across multiple countries. The global electrical grid — every power station, every transmission line, every transformer — operates on principles he established. When he built something, it worked. Not most of the time. Every time.

That last point is the one this book keeps coming back to, because it is the one that almost every other account of Tesla's life either overlooks or fails to grasp the significance of. Tesla was not a trial-and-error inventor. He did not build prototypes to see what would happen. He worked things out completely in his mind — sometimes in extraordinary visual detail, down to the dimensions of individual components — and then built them once. And they worked. The rotating magnetic field worked. The polyphase system worked. Niagara Falls worked. The Tesla coil worked. The magnifying transmitter worked. Colorado Springs worked. His track record across four decades of engineering, at every scale from a tabletop apparatus to the continental power grid, was perfect.

This matters because of what came next. After proving every principle at every available scale, Tesla said he could apply those same principles to the planet itself — that the earth could be set resonating at its natural frequency, and that electrical energy could be transmitted through its body to any point on its surface. He designed a machine to do it. He began building it. And then the funding collapsed, the tower was demolished, and Tesla spent the last four decades of his life insisting that the physics was proven and only the money had failed.

Most books treat this as tragedy, or hubris, or delusion. This book treats it as an open question — and not the kind of open question that politely acknowledges both sides. When a man with a flawless engineering record says he can do something, and every smaller-scale test of the underlying principles has confirmed his predictions, the burden of proof does not rest on him. It rests on everyone who assumes he was wrong without bothering to check.

* * *

A word about method. This is not a comprehensive biography. Tesla's life has been thoroughly documented by others — by John J. O'Neill, by Marc Seifer, by W. Bernard Carlson, and by many more. This book does not attempt to duplicate their scholarship. What it attempts is something different: to follow a single thread through Tesla's career, from the first insight to the last, and to take that thread seriously on its own terms.

The thread is resonance.

Every major invention Tesla produced was an application of the same principle: find the natural frequency of a system, drive it in rhythm, and the system amplifies your input beyond all proportion to the energy supplied. He discovered this principle in his twenties, applied it to electric motors, then to power transmission, then to high-frequency circuits, then to wireless energy transfer, and finally to the earth itself. At each stage, the principle worked. At each stage, the scale increased. The question this book asks is simple: is there any reason, other than money, that the final stage would not have worked too?

* * *

This book is also, deliberately, written with respect. Not reverence — Tesla's failures are described honestly throughout, and there were many. He was a dreadful businessman. He managed relationships poorly. He concealed the true scope of his plans from investors. He made enemies he didn't need to make and lost allies he couldn't afford to lose. His later positions on relativity, quantum mechanics, and the structure of the atom were behind their time, not ahead of it. His insistence on the ether is routinely cited as proof that he was stuck in the nineteenth century, though it is worth noting that Tesla's ether was not the rigid luminiferous medium that Michelson-Morley disproved — he described something dynamic, fluid-like, a medium from which matter itself arises as vortices of motion, and modern physics has quietly arrived at a quantum vacuum that is not empty either. None of this is offered as an excuse. It is offered as context for a man whose theoretical vocabulary lagged behind his engineering intuition, and whose direct experience with electromagnetic phenomena at scales nobody else had worked at gave him information his language could not fully express.

But respect means something specific here. It means not treating Tesla's greatest claim — that the earth can be used as a medium for wireless power transmission — as a fantasy to be indulged or a delusion to be diagnosed. It means looking at the engineering record, acknowledging that every prior application of these principles produced exactly the result Tesla predicted, and then sitting with the uncomfortable possibility that a man the world wrote off as a dreamer may have been telling the truth about the biggest thing he ever attempted.

That possibility is what this book is about.

CHAPTER ONE

THE BOY WHO SAW MACHINES

Smiljan and Gospić, 1856–1875

On a dry winter evening in the village of Smiljan, in what is now Croatia, a boy stroked his cat and the world lit up.

> *"As I stroked Macak's back," Nikola Tesla would remember eight decades later, "I saw a miracle that made me speechless with amazement. Macak's back was a sheet of light and my hand produced a shower of sparks loud enough to be heard all over the house."*

He asked his father what caused it. Milutin Tesla, an Orthodox priest, thought for a moment and said: "Well, this is nothing but electricity, the same thing you see through the trees in a storm."

The boy kept stroking. The sparks kept coming. And a question formed that would drive the next seventy years of his life: "Is nature a gigantic cat? If so, who strokes its back?"

It is a child's question, and Tesla knew that when he told the story. But it is also a physicist's question, and that is the part most people miss. The five-year-old in Smiljan was not reaching for a metaphor. He was observing a physical phenomenon — static discharge in dry air — and asking what it told him about the nature of the force that produced it. He wanted to know the mechanism. He wanted to know what electricity actually *is*. "Eighty years have gone by since that time," he wrote in 1939, "and I still ask the same question, unable to answer it."

This book is about what happened between the question and the admission. It is about a man who spent his life not just harnessing electricity but trying to understand what it is — and who built some of the most consequential machines in human history as experimental tests of his answer.

But the ideas did not come from nowhere. They came from a particular place, a particular family, and a particular mind — one that worked in ways so unusual that Tesla himself struggled to explain it.

* * *

Smiljan sits in the Lika region of the Dinaric Alps, a landscape of limestone, running water, and hard weather. Tesla's ancestors had migrated there from western Serbia in the 1690s, part of the great displacement of Christian populations by the Ottoman Turks. The soil was so rocky that the Likan Serbs had a saying: when God was distributing rocks over the Earth, he carried them in a sack, and when he was above their land, the sack broke.

The Tesla family — the name means "adze," a small woodcutting tool — had pursued military and clerical careers for generations. Tesla's grandfather Nikola served in Napoleon's army. His uncle Josif became a professor of mathematics at a military academy. His father Milutin trained as an officer but found military life intolerable (a reprimand for unpolished brass buttons was the final straw) and became a priest instead. He married Djuka Mandic, the daughter of another priest, from a family in which virtually every man entered either the clergy or the military.

Milutin was a reformer and an intellectual. He assembled a library of books in several languages — religion, mathematics, science, literature — and could recite whole volumes of poetry from memory. He wrote for Serbian newspapers, argued publicly for schools where Serbs could be taught in their own language, and preached with enough fire to earn the Red Sash from his bishop. He was also, by his son's account, a man of formidable mental discipline who drilled young Nikola with exercises in memory, mental calculation, and critical reasoning.

Tesla's mother, Djuka, was something else entirely. She was an inventor.

> *My mother was an inventor of the first order and would, I believe, have achieved great things had she not been so remote from modern life and its multifold opportunities. She invented and constructed all kinds of tools and devices and wove the finest designs from thread which was spun by her. When she was past sixty, her fingers were still nimble enough to tie three knots in an eyelash.*

This detail — three knots in an eyelash — has been quoted so often that it has become a biographical cliché. But think about what it actually means. Djuka Tesla was a woman of extraordinary manual precision, living in a remote mountain village, surrounded by peasants using crude tools unchanged for centuries, and she spent her life designing and building better ones. She worked from four in the morning until eleven at night, every day, and most of the tools, furnishings, and clothing in the household were products of her hands and her ingenuity.

This is the household in which Nikola Tesla learned what it meant to make things: a father who could hold entire books in his memory and who drilled his children in mental calculation, and a mother who designed and built machines from whatever materials the mountains provided. The intellectual framework came from one parent. The instinct to construct came from the other.

* * *

And then there was the mind itself — which was, by Tesla's own account, extraordinary and frightening in equal measure.

> *In my boyhood I suffered from a peculiar affliction due to the appearance of images, often accompanied by strong flashes of light, which marred the sight of real objects and interfered with my thought and action. They were pictures of things and scenes which I had really seen, never of those I imagined. When a word was spoken to me the image of the object it designated would present itself vividly to my vision and sometimes I was quite unable to distinguish whether what I saw was tangible or not.*

Read that passage carefully. Tesla is not describing a vivid imagination. He is describing involuntary hallucination-grade imagery so intense that he could not reliably tell it apart from the physical world. A word spoken aloud would conjure the corresponding image with such force that it occluded his vision of real objects. A disturbing scene witnessed during the day would replay before his eyes at night, fixed in space, persisting even when he pushed his hand through it.

Tesla consulted psychologists and physiologists throughout his life. None of them could explain it. His own theory was that the images were "the result of a reflex action from the brain on the retina under great excitation." Modern neuroscience would probably recognise something in the spectrum of eidetic imagery or hypnagogic hallucination — a visual system wired so tightly that internal representations could override sensory input. Whatever the mechanism, the experience was real and it was tormenting.

And then he did something remarkable with it.

To escape the images, the young Tesla tried concentrating on other things he had seen. This worked briefly, but he soon ran through everything in his limited experience — "my 'reel' had run out." So he began generating new images deliberately, scenes of places he had never visited, and found that if he let his imagination range freely, the tormenting visions receded. He began to travel in his mind — seeing cities, meeting people, having experiences as vivid and emotionally real as anything in his waking life. He did this every night for years.

By the time he was seventeen, this coping mechanism had evolved into something far more powerful: a complete design methodology.

> *I observed to my delight that I could visualize with the greatest facility. I needed no models, drawings or experiments. I could picture them all as real in my mind. Thus I have been led unconsciously to evolve what I consider a new method of materializing inventive concepts and ideas, which is radically opposite to the purely experimental and is in my opinion ever so much more expeditious and efficient.*

The passage that follows is one of the most extraordinary claims any inventor has ever made, and it deserves to be quoted at length, because the entire argument of this book depends on taking it seriously:

> *My method is different. I do not rush into actual work. When I get an idea I start at once building it up in my imagination. I change the construction, make improvements and operate the device in my mind. It is absolutely immaterial to me whether I run my turbine in thought or test it in my shop. I even note if it is out of balance. There is no difference whatever, the results are the same. In this way I am able to rapidly develop and perfect a conception without touching anything. When I have gone so far as to embody in the invention every possible*

> *improvement I can think of and see no fault anywhere, I put into concrete form this final product of my brain. Invariably my device works as I conceived that it should, and the experiment comes out exactly as I planned it. In twenty years there has not been a single exception.*

There it is. Not "In twenty years I have had a good success rate." Not "Usually my designs work more or less as expected." He says: in twenty years, not a single exception. The machine built in his mind and the machine built in his shop produce identical results. Every time.

This is either the most grandiose self-delusion in the history of invention, or it is a factual report from a man whose visual-constructive abilities were so far beyond the normal range that they constituted a different kind of engineering. The rest of Tesla's career — the rotating magnetic field conceived in a park and drawn in the sand, the AC motor that worked exactly as imagined on the first try, the Tesla coil, the Colorado Springs transmitter — suggests that it was the latter. And it all grew from a childhood affliction that he learned, by an act of will, to turn into a tool.

* * *

While this inner transformation was taking place, the outer boy was building things. And the things he built reveal, in miniature, every principle that would later drive his life's work.

His first serious invention was a rotary engine powered by June bugs. He attached four of them to a crosspiece mounted on a spindle, connected the spindle to a larger disc, and let the insects drive the whole assembly. "These creatures were remarkably efficient," he noted with dry satisfaction, "for once they were started they had no sense to stop and continued whirling for hours and hours and the hotter it was the harder they worked." The project ended when a neighbourhood boy ate the bugs alive, and Tesla could never touch an insect again. But look at what the machine was: a biological rotary engine — natural energy harnessed to continuous rotation. The boy instinctively reached for the same solution he would reach for at twenty-six in Budapest, at thirty-two at Niagara Falls, and at forty-three in Colorado Springs: find a source of energy, convert it to rotary motion, and extract useful work.

He took apart his grandfather's clocks ("in the former operation I was always successful but often failed in the latter"). He built pop-guns from hollow stalks, using compressed air to fire hemp plugs. He carved wooden swords and mowed down imaginary enemies in his mother's cornfield. And then, in the schoolroom at Gospić, he found demonstration models of waterwheels and turbines, and something clicked.

> *I constructed many of these and found great pleasure in operating them. I was fascinated by a description of Niagara Falls I had perused, and pictured in my imagination a big wheel run by the Falls. I told my uncle that I would go to America and carry out this scheme. Thirty years later I saw my ideas carried out at Niagara and marveled at the unfathomable mystery of the mind.*

There are two things to notice about this passage. The first is the sheer improbability of a ten-year-old Serbian boy in a mountain village imagining himself harnessing Niagara Falls — and then doing it. The second, and for our purposes more important, is what it tells us about how Tesla's mind worked. He read a description of Niagara Falls. He "pictured in his imagination" a wheel run by the falls. The image was so real and so persistent that it survived thirty years and an ocean crossing. When the Niagara power plant was finally built, using Tesla's polyphase AC system, it was not a coincidence. It was the completion of a visualization that had begun in a schoolroom in Lika.

* * *

Two episodes from Tesla's childhood in Gospić are worth lingering over, because they show something that the later biographical tradition tends to obscure: Tesla was not just a dreamer. He was someone whose theoretical understanding manifested as immediate physical competence under pressure.

The first is the fire pump. An enterprising young merchant in Gospić had organized a volunteer fire department, complete with uniforms and a handsome new engine painted red and black. The day of the official demonstration arrived. The entire town turned out. Sixteen men pumped furiously. No water came. The experts examined the machine. Nothing worked. The spectacle was a complete embarrassment.

Tesla was watching from the crowd. He walked to the river, felt for the suction hose, and found that it had collapsed. He waded in, opened it up, and water gushed from the nozzle. He was carried home on the firemen's shoulders.

He tells this story with characteristic understatement — "my knowledge of the mechanism was nil and I knew next to nothing of air pressure, but instinctively I felt for the suction hose" — but what it actually demonstrates is a child who could diagnose a pneumatic failure in an unfamiliar machine by reasoning from first principles. No one told him to check the intake. No one taught him about suction collapse. He understood, at some intuitive level, that a pump that won't draw must have a blockage on the input side, and he acted on that understanding while the adults stood around looking at each other.

The second episode is more dramatic, and more revealing.

Near Gospić there was a flour mill with a dam across the river. Tesla, who was a strong swimmer, used to swim out to the dam for sport. One day he went alone, and found that the water had risen and was sweeping him toward the edge. He grabbed the wall with both hands, but the pressure against his chest was enormous, and he could barely keep his head above the surface. No one was in sight. His voice was lost in the roar of the waterfall.

> *Just as I was about to let go, to be dashed against the rocks below, I saw in a flash of light a familiar diagram illustrating the hydraulic principle that the pressure of a fluid in motion is proportionate to the area exposed, and automatically I turned on my left side. As if by magic the pressure was reduced and I found it comparatively easy in that position to resist the force of the stream.*

He then worked his way slowly along the dam face, left-handed (he was ambidextrous later but left-dominant as a child), and reached the bank half-dead, with the skin torn from his left side.

The detail that matters is the flash of light. Not a memory. Not a gradual recollection. A visual image of a hydraulic diagram — the principle that pressure is proportional to exposed area — appearing unbidden in the moment of extremity and immediately translating into a physical action that saved his life. This is the same visual system that produced the tormenting images, the same one that would later allow him to build machines in his mind and note whether they were out of balance. Under the pressure of the water and the proximity of death, it delivered a piece of theoretical physics as a flash of light, and his body responded automatically.

If you are inclined to dismiss Tesla's later claims about the vividness and reliability of his mental imagery, remember this episode. The mental image worked. It produced an immediate, correct, life-saving physical response. The boy on the dam wall is the same man who will draw the rotating magnetic field in the sand in Budapest, having seen it complete in his mind. The mechanism is the same. Only the scale changes.

* * *

There was one more development in Tesla's childhood that deserves attention, because it reveals the philosophical bent that would later separate him from virtually every other electrical engineer of his era.

As he learned to master his visual affliction, Tesla noticed something: the images that appeared were always triggered by something he had actually seen. They were not random. They were responses to external stimuli. He began tracking the chain of cause and effect — this image was triggered by that sight, that thought was prompted by this impression — and what he found unsettled him profoundly.

> *Soon I became aware, to my surprise, that every thought I conceived was suggested by an external impression. Not only this but all my actions were prompted in a similar way. In the course of time it became perfectly evident to me that I was merely an automaton endowed with power of movement, responding to the stimuli of the sense organs and thinking and acting accordingly.*

A teenage boy in a mountain village in the Austro-Hungarian Empire, the son of an Orthodox priest, arrived independently at a mechanistic philosophy of mind. Not through reading Descartes or La Mettrie — through direct introspective observation of his own cognitive processes. He watched his own thoughts arise, traced them to their sensory causes, and concluded that he was a machine. A "meat machine," as he would later put it.

One can only imagine how this went over with his father.

But for the argument of this book, the automaton insight is the first seed of Tesla's theoretical framework. If a human being is a mechanical system responding to external forces, then why not the universe? If thought is the result of physical impressions transmitted through a medium (the nervous system), then why not electricity? If the boy on the dam wall is a machine that responds to hydraulic pressure by orienting its body to minimize exposed area, then the boy has already understood, in his own flesh, the principle that forces are transmitted through media and that the geometry of the response matters.

Tesla would spend the rest of his life exploring this idea: that electricity, magnetism, light, and gravity are all mechanical phenomena — stresses and motions transmitted through a physical medium. His inventions would be designed to test that idea. His experiments would be structured to reveal that medium's behavior. And it all begins here, in a boy watching water, stroking a cat, building bug-powered turbines, and noticing that his own mind operates by the same mechanical principles as the world outside it.

* * *

By the time he left Gospić for the Higher Gymnasium at Karlovac at the age of fourteen, the essential Tesla was already formed. The visual imagination that would allow him to design machines without drawings or models. The habit of thinking in terms of mechanical forces — pressure, flow, vibration. The conviction that theory should precede construction. The ability to translate theoretical understanding into immediate physical action.

And one thing more, perhaps the most important: an instinct for resonance.

Every child who has ever been pushed on a swing understands the principle before they have a word for it. Push at the right moment — matching the swing's natural rhythm — and the arc grows with almost no effort. Push at the wrong moment and everything fights you. The timing matters more than the force. The boy who turned sideways against the dam wall understood this in his bones. The boy who freed the fire pump understood it in his hands. The boy who built a rotary engine from June bugs and dreamed of harnessing Niagara Falls understood it as a vision that would not let him go.

Resonance — the discovery that small forces, precisely timed, can produce enormous effects — would become the governing principle of Tesla's life's work. The rotating magnetic field, the polyphase motor, the Tesla coil, the mechanical oscillator, wireless power transmission: every one is a machine built to find a natural frequency and match it. He did not overpower systems. He listened for their rhythm and aligned with it. This is what set him apart from Edison, who tried ten thousand filaments by brute experiment, and from the imitators who would later transmit his signals using his own apparatus without understanding what carried them. Tesla worked from the principle outward. He believed — and spent his life demonstrating — that if you understood the medium, the medium would do the work for you.

This book follows that thread. It is not a comprehensive biography. It is the story of one idea — resonance as the master key — traced through the life of the man who pursued it further than anyone before or since. The machines are here. So are the failures, the feuds, the financial disasters, and the long decline. But the through-line is always the same: a boy in a mountain village who noticed that the world responds to rhythm, and who never stopped asking what that meant.

What he did not yet have was a problem worthy of these gifts. That would come at Graz, where a professor would spin up a Gramme dynamo in a lecture theatre and Tesla would watch the commutator spark and think: there must be a better way.

The way would take six years to find. It would come to him in a park in Budapest, at sunset, while reciting Goethe. And it would change the world.

CHAPTER TWO

THE SPARK AND THE COMMUTATOR

Graz, Prague, and Budapest, 1875–1882

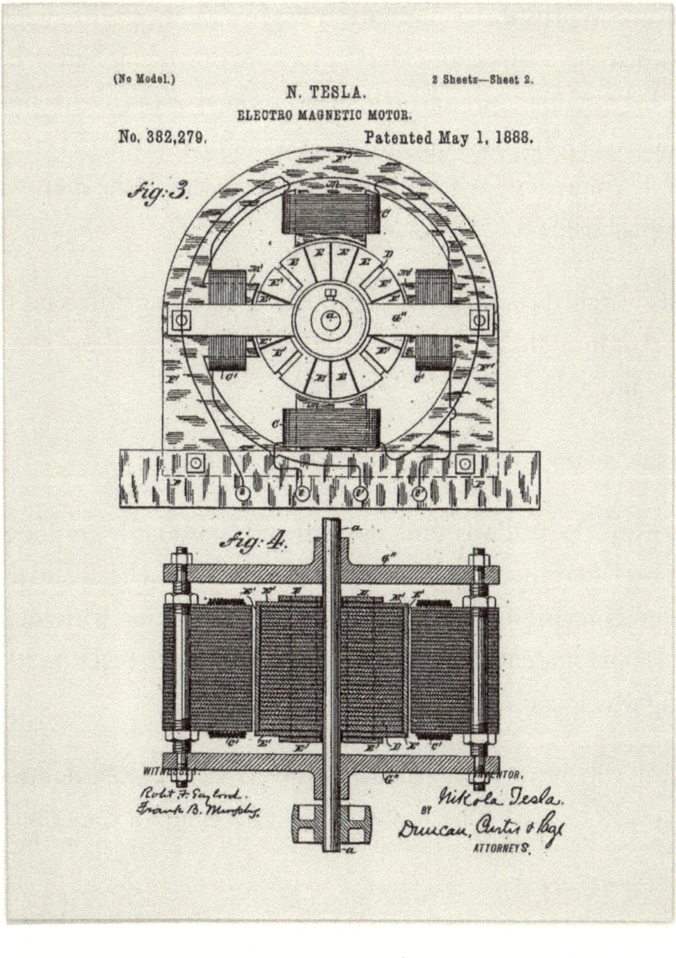

The machine arrived from Paris in a wooden crate sometime during Tesla's second year at the Joanneum Polytechnic in Graz. It was a Gramme dynamo — a horseshoe-shaped laminated field magnet with a wire-wound armature and a commutator — and Professor Poeschl connected it up in the lecture theatre and ran it as a motor to demonstrate the effects of electrical currents.

The brushes sparked.

This wasn't unusual. Sparking was the price you paid for direct current. The commutator — a mechanical switch that reversed the direction of current flow in the armature coils twice per revolution — required carbon brushes to maintain contact with its rotating segments, and those brushes scraped and arced and wore down and had to be replaced. Every DC motor and generator in the world had this problem. It was considered an engineering nuisance, not a theoretical deficiency. You managed it. You redesigned the brushes. You accepted the sparks as the cost of doing business.

Tesla watched the sparks and thought something different. He observed that it might be possible to operate a motor without these appliances altogether.

He said so out loud.

Professor Poeschl was a methodical and thoroughly grounded German, with enormous feet, hands like the paws of a bear, and a gift for experimental precision. He listened to Tesla's suggestion with interest. Then he did his student the honor of devoting an entire lecture to explaining why it couldn't be done.

The argument was simple and, by the physics of 1877, devastating. A motor works by applying a force to make a conductor rotate inside a magnetic field. Direct current flows in one direction; the commutator reverses the current in each coil at exactly the right moment to keep the rotation going. Without the commutator, the current would not reverse, the force would not sustain itself, and the motor would stop. Eliminating the commutator was therefore "equivalent to converting a steadily pulling force, like that of gravity, into a rotary effort. It is a perpetual motion scheme, an impossible idea."

At the conclusion of his lecture, Poeschl added a remark that Tesla would carry with him for the rest of his life: "Mr. Tesla may accomplish great things, but he certainly never will do this."

* * *

To understand what happened next — and what it cost — you need to understand what Tesla had actually seen in those sparks.

Poeschl's argument was logically correct within the framework of direct current. If the current flows in one direction, you need a mechanical switch to reverse it. That's what the commutator does, and without it, the motor fails. This wasn't wrong. It was something more dangerous than wrong: it was a correct answer to the wrong question.

The question Poeschl was answering was: how do you build a DC motor without a commutator? The answer is: you can't. But that wasn't the question Tesla was asking. Tesla was asking: why does the current need to be reversed mechanically when nature already alternates it?

An electrical generator, by the physics of electromagnetic induction, naturally produces alternating current. The conductor moves through the magnetic field, the induced voltage rises and falls and reverses direction with each half-turn of the rotor. This is what generators *do*. The alternation isn't a defect. It's the natural output of a rotating conductor in a magnetic field. The commutator exists to fight this alternation — to chop and redirect it into a one-way flow that the rest of the circuit can use. Every spark at the brushes is the sound of a machine being forced to behave against its own nature.

What Tesla saw in the sparking Gramme dynamo was a machine at war with itself. And his instinct — the same instinct that had sent him wading into the river to open the collapsed suction hose, the same instinct that had turned him sideways against the dam wall — was that the war was unnecessary. If the current wanted to alternate, then build a motor that *uses* the alternation. Don't fight the rhythm. Match it.

He didn't yet know how. But he knew, with the certainty of intuition that transcends logic, that the answer existed. "Instinct is something which transcends knowledge," he wrote. "We have, undoubtedly, certain finer fibers that enable us to perceive truths when logical deduction, or any other willful effort of the brain, is futile."

Poeschl's rebuttal shook him. The professor's authority was formidable, his logic was airtight within its premises, and Tesla wavered. But not for long. "Soon I became convinced I was right," he wrote, "and undertook the task with all the fire and boundless confidence of youth."

The task would consume the next six years of his life and very nearly destroy him.

* * *

Tesla's method of attack was entirely characteristic. He didn't go to the workshop. He went into his own mind.

I started by first picturing in my mind a direct-current machine, running it and following the changing flow of the currents in the armature. Then I would imagine an alternator and investigate the processes taking place in a similar manner. Next I would visualize systems comprising motors and generators and operate them in various ways. The images I saw were to me perfectly real and tangible.

This is the design methodology he had described in the previous chapter — the visual system that had evolved from his childhood affliction. He was now applying it to the hardest problem in electrical engineering, and it worked exactly as he said it did, except that in this case it didn't produce a solution. He could see the machines. He could run them in his mind, follow the currents, observe the forces. But the key insight — how to create a rotating magnetic field without mechanical commutation — wouldn't come.

"All my remaining term in Gratz was passed in intense but fruitless efforts of this kind," he wrote, "and I almost came to the conclusion that the problem was insolvable."

Notice what's happening. A twenty-year-old student is spending his evenings running imaginary electromagnetic systems in his head, visualizing alternating currents, mentally constructing and testing motor after motor, and finding that none of them work. The images are perfectly real and tangible. The failures are perfectly real and tangible. He's living inside the problem, and the problem isn't yielding.

This is the other side of Tesla's extraordinary visual gift. When it delivered an answer — as it had on the dam wall, as it would in Budapest — the answer came with absolute clarity and conviction. But when it didn't deliver, there was no fallback position. He couldn't sketch a rough approximation and test it in the shop. He couldn't build a prototype and iterate toward a solution. His method was all or nothing: either the complete machine appeared in his mind, correct in every detail, or there was nothing.

For six years, there was nothing.

* * *

Tesla's first year at Graz had been a performance of almost inhuman discipline. He rose at three in the morning and worked until eleven at night, no Sundays or holidays excepted. He passed nine examinations — nearly double the required number — and his professors wrote to his father that unless the boy was removed from the institution, he would kill himself with overwork. When Tesla went home expecting a triumph, his father made light of the achievement. Tesla was mortified. Only years later, after Milutin's death, did he find the letters.

The second year was the year of the Gramme dynamo and the confrontation with Poeschl. It was also the year Tesla lost his scholarship. Whether the two events were connected is unclear; what is certain is that without the scholarship and carrying the unsolvable problem like a weight in his chest, Tesla's relationship with the institution began to fracture.

He discovered card games. Then billiards. Then he discovered that he couldn't stop.

> *At a certain age I contracted a mania for gambling which greatly worried my parents. To sit down to a game of cards was for me the quintessence of pleasure.*

The word *mania* is precise. Tesla threw himself into gambling with the same intensity he had applied to his studies — sessions lasting forty-eight hours at a stretch, his entire allowance and tuition money wagered and lost and sometimes won back and lost again. His father, the priest who led an exemplary life, couldn't excuse the senseless waste and said so repeatedly. Tesla's defense was characteristic: "I would say to him, 'I can stop whenever I please but is it worth while to give up that which I would purchase with the joys of Paradise?'"

It's a strange thing to watch, this collapse. The boy who rose at three and worked until eleven, who eclipsed all records, who carried nine examinations in a single year, was now spending days at the card table while the unsolved problem of the commutator burned somewhere beneath the compulsion. Biographers have struggled with it. Some blame the loss of the scholarship. Some blame the confrontation with Poeschl. Some see the gambling as a displacement activity — a mind built for obsessive focus redirecting itself when the primary obsession wouldn't yield.

There may be something simpler at work. Tesla's visual system was running electromagnetic machines all day and all night and producing nothing but failures. Forty-eight hours at the billiard table may have been the only thing that could make it stop.

Whatever the cause, the result was unambiguous. By the third year Tesla had stopped attending lectures. He wasn't registered for the following spring. He never sat his final examinations and he never received a degree from the Joanneum Polytechnic.

His mother saved him, with a gesture of such psychological acuity that it bears repeating. One afternoon, having lost everything, Tesla was craving another game. Djuka came to him with a roll of bills and said:

"Go and enjoy yourself. The sooner you lose all we possess the better it will be. I know that you will get over it."

She was right. He conquered the passion, as he put it, "then and there," and not only vanquished it but tore it from his heart so as not to leave even a trace of desire. It was the same act of will that had tamed the tormenting images in childhood. Tesla's relationship with his own compulsions was binary: total surrender or total eradication. There was no moderation in him. There never would be.

* * *

In December 1878, at the age of twenty-two, Tesla left Graz and severed all contact with his family.

The shame of having dropped out — the son of a priest, the scholarship student, the boy whose professors wrote home about his brilliance — was more than he could face. He moved to Maribor, in what is now Slovenia, took a job as a draftsman at an engineering firm for sixty florins a month, and spent his free time playing cards in the streets with locals. When his father tracked him down and tried to bring him home, Tesla refused. Eventually the police settled the matter: he was escorted from Maribor to Gospić for lacking a residence permit.

He returned to his family in disgrace. And then, in 1879, Milutin Tesla died.

Tesla rarely spoke of this period. His autobiography skips from the gambling to the nervous breakdown with barely a connecting sentence, as if the years between Graz and Budapest were a void he preferred not to illuminate. We know that after his father's death he taught for a time in Gospić. We know that he spent a brief period at the University of

Prague, where he made what he called "a decided advance, which consisted in detaching the commutator from the machine and studying the phenomena in this new aspect, but still without result." We know that the problem was still alive in him, unsolved and unrelenting.

And we know that in 1881, he moved to Budapest to take a position with the new telephone exchange being installed there by a friend of his family, and that almost immediately upon arrival, his mind and body collapsed.

* * *

"What I experienced during the period of that illness surpasses all belief."

Tesla's nervous breakdown in Budapest is one of the strangest medical episodes in the history of science, and he described it with the clinical precision of a man determined to leave a record even of his own disintegration.

His senses, always acute, became catastrophically amplified. He could hear the ticking of a watch with three rooms between himself and the timepiece. A fly alighting on a table caused a dull thud in his ear. A carriage passing at a distance of several miles shook his entire body. The vibration of the ground beneath his feet, transmitted through a chair, was painful. Sunlight falling on his face felt like a physical blow. A rubber cushion placed beneath his bed frame had to be changed constantly because the pressure of his own body produced intolerable sensations. He could feel his pulse and the movement of his internal organs, and the experience was agonizing.

He was, in the most literal sense, resonating with everything. Every vibration in his environment — acoustic, mechanical, optical —

was being received and amplified by a nervous system that had lost the ability to attenuate. The same perceptual system that could visualize electromagnetic machines with hallucinatory clarity was now receiving the physical world at full gain, without filtering, without damping, without the normal physiological mechanisms that protect consciousness from the overwhelming totality of sensory input.

His doctors gave him up. His friends expected him to die.

There's a temptation to see this breakdown as a simple case of overwork and nervous exhaustion, and that's part of the truth. Tesla had been driving himself mercilessly since the age of fourteen. He had endured cholera, malaria, the death of his father, the humiliation of Graz, the unsolvable problem of the commutator, and the accumulated strain of a perceptual system that was never entirely under his control. Any one of these might have broken a lesser constitution.

But there's something else in the breakdown that deserves attention, because it connects directly to the argument of this book. Tesla's amplified senses weren't producing random noise. They were producing signal. He was hearing real sounds, feeling real vibrations, perceiving real phenomena — just at a scale of sensitivity that normal human physiology can't sustain. The watch was actually ticking. The fly was actually landing. The carriage was actually passing. His nervous system had become, for a period of weeks, an instrument of extraordinary fidelity and no selectivity whatsoever. He was receiving everything, on every frequency, all at once.

This is what happens to a resonant system with no damping. The amplitude builds without limit until the structure fails. Tesla's body was the structure, and it very nearly failed.

* * *

The cure, when it came, was remarkably simple. His friend Antal Szigeti — a fellow engineer and one of the few people who stayed close to Tesla during this period — got him outside and walking.

They walked in the Városliget, the great city park of Budapest, every day. Fresh air, physical movement, the gradual re-engagement of the body with the rhythms of ordinary life. Tesla's superhuman sensitivity began to recede. His hearing returned to merely extraordinary. His pulse stopped being an event. The world, which had been an assault, became navigable again.

He took a position at the Budapest telephone exchange and threw himself into the work. He made several improvements to the central station apparatus and designed a telephone repeater, or amplifier — a device that wouldn't be built commercially for years. His mind was functioning again, and with it, the old problem returned. The commutator. The spark. The alternating current that wanted to alternate and wasn't being allowed to.

Tesla had now been carrying this problem for six years. He had attacked it mentally at Graz, advanced it theoretically at Prague, and nearly been destroyed by it in Budapest. He had tried every configuration he could imagine and none of them worked. The conventional wisdom, backed by Poeschl's authority and the entire tradition of DC electrical engineering, said it couldn't be done.

And then, on a February afternoon in 1882, walking in the Városliget with Szigeti, watching the sun go down, he began to recite poetry.

* * *

Tesla knew entire books by heart — an inheritance from his father, who could recite whole volumes from memory. One of these was Goethe's *Faust*. The sun was setting over the park, and it reminded him of a passage from the end of the first act, in which Faust watches the sun disappear and yearns to follow it:

> *The glow retreats, done is the day of toil;*
> *It yonder hastes, new fields of life exploring;*
> *Ah, that no wing can lift me from the soil,*
> *Upon its track to follow, follow soaring!*

Notice the rhythm of the verse itself. It alternates — between longing and motion, between earth and sky, between the glow that retreats and the fields it hastes toward. The lines rock back and forth like a pendulum, like a swing, like an alternating current. Tesla's mind had been saturated with the problem of alternation for six years. He was reciting a poem whose rhythm was alternation, while watching the sun trace an arc across the sky. The resonance between the verse and the problem may have been unconscious. But Tesla's mind was not a mind that missed resonance.

What happened next is the most famous moment in the history of electrical engineering, and Tesla told the story the same way every time he told it, for over fifty years, with the unwavering conviction of a man describing the central event of his life.

> *As I uttered these inspiring words the idea came like a flash of lightning and in an instant the truth was revealed. I drew with a stick on the sand the diagrams shown six years later in my address before the American Institute of Electrical Engineers, and my companion understood them perfectly. The images I saw were wonderfully sharp and*

clear and had the solidity of metal and stone, so much so that I told him: "See my motor here; watch me reverse it."

There it is. Six years of fruitless effort, a complete nervous breakdown, recovery, and then — while reciting Goethe at sunset — the entire solution appears at once, complete in every detail, with the solidity of metal and stone.

What did he see?

He saw the rotating magnetic field.

The principle is this. Instead of using a single alternating current to drive a motor — which, as Poeschl correctly observed, would produce a force that merely pulsed back and forth without sustaining rotation — you use two or more alternating currents that are out of phase with each other. Each current drives its own set of coils in the stator, the stationary housing of the motor. Because the currents peak at different times, the magnetic field produced by the stator doesn't pulse. It rotates. It sweeps around the inside of the motor housing like a beam of light from a lighthouse, and the rotor — the moving part — follows it, dragged along by electromagnetic induction.

No commutator. No brushes. No sparks. No mechanical switching of any kind. The alternation that the commutator had been fighting was now the very thing that made the motor work. The current alternated; the coils were arranged so that the alternation produced rotation; and the rotor turned, silently and smoothly, driven by nothing but the natural rhythm of the current itself.

It was, in the language of this book, a perfect resonance solution. Tesla had stopped fighting the medium and started working with it. The alternating current wasn't a problem to be corrected. It was a feature to be exploited. The natural frequency of the electrical system — the oscillation that every generator produced as a matter of physics — was the very thing that, properly arranged, would create continuous rotary motion without any mechanical intervention at all.

Poeschl had been right that you can't convert a steadily pulling force into rotation without a commutator. But the force wasn't steady. It was alternating. It was *already* oscillating, already pulsing, already going back and forth. Tesla's insight was that if you stopped trying to straighten the oscillation into a steady pull and instead *used* the oscillation — if you arranged the geometry so that multiple oscillations, staggered in time, produced a rotating field — you could build a motor that was simpler, more efficient, and more powerful than anything the direct-current world could produce.

The child on the swing doesn't fight the swing's natural period. The engineer of the rotating field doesn't fight the current's natural alternation. In both cases, the answer's the same: match the rhythm, and the medium does the work.

* * *

Tesla's description of what followed is ecstatic.

> *I cannot begin to describe my emotions. Pygmalion seeing his statue come to life could not have been more deeply moved. A thousand secrets of nature which I might have stumbled upon accidentally I would have given for that one which I had wrested from her against all odds and at the peril of my existence.*

He spent the next two months in a state of continuous creative exaltation, evolving in his mind virtually every type of motor and modification of the polyphase system that would later be identified with his name. The machines came one after another, complete and operational, running in his imagination with the same reliability they would later demonstrate in the physical world. He could see the motors spinning. He could reverse them. He could watch the rotating field sweep around the stator and feel the rotor follow.

> *It was a mental state of happiness about as complete as I have ever known in life. Ideas came in an uninterrupted stream and the only difficulty I had was to hold them fast. The pieces of apparatus I conceived were to me absolutely real and tangible in every detail, even to the minute marks and signs of wear. I delighted in imagining the motors constantly running, for in this way they presented to the mind's eye a more fascinating sight.*

This passage is worth pausing over, because it reveals something crucial about how the discovery arrived. Tesla didn't reason his way to the rotating magnetic field through a chain of logical deductions. He didn't derive it mathematically and then verify it experimentally. He saw it. The complete machine appeared in his visual field with the solidity of metal and stone — the same visual system that had tormented him as a child, the same one that had shown him the hydraulic diagram on the dam wall, the same one that had been running and failing to solve the problem for six years. When the answer finally came, it came not as an equation or an argument but as an image: a motor, running, reversible, real.

The six years of failure weren't wasted. They were the problem being loaded into a system that required everything to be in place before it could produce an output. Every configuration Tesla had visualized and rejected at Graz, every theoretical advance he had made at Prague — "detaching the commutator from the machine and studying the phenomena in this new aspect" — was part of the loading. The breakdown in Budapest, however terrible, may have been part of it too: a perceptual system pushed to its absolute limit, burning through every filter, before reassembling itself around the answer it had been seeking.

Or perhaps not. Perhaps the connection between the breakdown and the vision is coincidental, and the answer simply arrived when it arrived, as solutions to long-meditated problems sometimes do. Tesla himself offered no causal explanation. He said the idea came like a flash of lightning, and he left it at that.

What's certain is that on a February afternoon in 1882, in a park in Budapest, while reciting Goethe at sunset, a twenty-five-year-old Serbian engineer who had no degree, no money, no workshop, and no reputation saw the complete design for a machine that would transform the modern world. He drew the diagrams in the sand with a stick. His friend understood them perfectly.

Now he had to build it. And to build it, he had to get to the one place in the world where electrical engineering was moving fast enough to keep up with what he had seen.

He had to get to America.

But first, he had to get to Paris.

CHAPTER THREE

THE ROTATING FIELD

Budapest, Paris, Strasbourg, and New York, 1882–1884

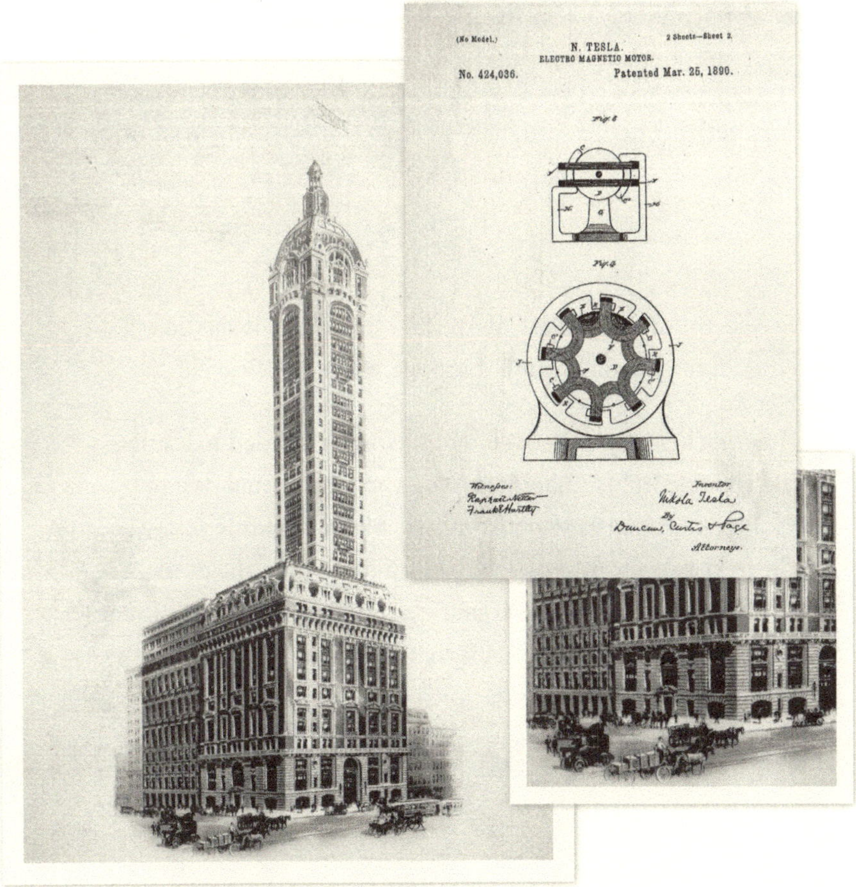

Tesla now had the most important electrical invention of the century complete in his head, and nowhere to put it.

This is the part of the story that doesn't get told often enough, because it's not dramatic in the way of sparking dynamos and visions in the park. It's the other kind of drama — the slow, grinding kind — in which a man with a world-changing idea walks from office to office, explains it clearly, demonstrates it convincingly, and is met with polite indifference. The rotating magnetic field was conceived in February 1882. Tesla wouldn't build a working motor in a proper laboratory until 1887. Five years. Five years in which the idea existed only in his mind and in diagrams scratched in sand and chalk, while he tried to find someone, anyone, who would give him the means to make it real.

The pattern will repeat throughout his life, and it begins here: Tesla invents something extraordinary, and then the world makes him wait.

* * *

In the weeks following the vision in the Városliget, Tesla did what he always did after a breakthrough. He went deeper.

The rotating field was a principle. Now he needed machines. Working entirely in his imagination — the same visual design methodology that had evolved from his childhood affliction — he spent the next two months evolving virtually every type of motor and modification of the polyphase system that would later bear his name. Single-phase motors, two-phase motors, three-phase motors. Motors with different winding configurations. Generators designed to produce polyphase current. Transformers to step the voltage up and down. The whole system, from generation to transmission to end use, assembled and tested in his mind before he'd touched a single piece of wire.

> *It was a mental state of happiness about as complete as I have ever known in life. Ideas came in an uninterrupted stream and the only difficulty I had was to hold them fast.*

Meanwhile, he still had a day job. The Budapest telephone exchange was operational and Tesla was its chief electrician, making improvements to the central station equipment and designing what he claimed was a telephone repeater or amplifier — a device that wouldn't be built commercially for years. When Tivadar Puskás, the family friend who'd arranged the position, asked how he was getting on, Tesla answered with characteristic precision: "The last twenty-nine days of the month are the toughest."

He was brilliant, employed, and broke. He had the design for the AC motor in his head, complete and ready to build. What he didn't have was a workshop, materials, funding, or anyone in Budapest who understood what he was talking about.

Puskás solved part of the problem. He had connections to the Continental Edison Company in Paris — Thomas Edison's European operation — and in 1882 he arranged for Tesla to join the firm. It wasn't a position that had anything to do with alternating current. Edison's entire empire ran on direct current, and any mention of AC was effectively taboo. But it was a job in the electrical industry, in one of the great cities of the world, and Tesla took it.

* * *

Paris in the early 1880s was electrifying itself. The Continental Edison Company was installing indoor incandescent lighting across the city, and Tesla was assigned to the Société Electrique Edison at Ivry-sur-Seine, on the southern outskirts, designing improvements to electrical equipment and troubleshooting DC power plants across France and Germany.

His daily routine was punishing and, characteristically, self-imposed. Every morning, regardless of weather, he walked from his lodgings on the Boulevard St. Marcel to a bathing house on the Seine, plunged into the water, swam twenty-seven laps, then walked an hour to the factory at Ivry. He arrived for what he called a "woodchopper's breakfast" at half past seven and spent the rest of the day solving problems for the plant manager, Charles Batchelor — Edison's intimate friend and right hand in Europe.

Twenty-seven laps. Not twenty-five, not thirty. Twenty-seven — divisible by three. The compulsions from childhood were still running.

Tesla impressed everyone he worked with. He was fast, thorough, and had an instinct for diagnosing electrical faults that bordered on the uncanny. The Americans at the plant, in particular, took a liking to him — partly for his engineering, partly because he turned out to be an exceptional billiards player. (The old Graz compulsion, it seems, hadn't been entirely wasted.) One of them, a foreman named D. Cunningham, offered to form a stock company around Tesla's AC invention. Tesla found the proposal "comical in the extreme" and didn't understand what a stock company was. Nothing came of it.

But the AC motor was always there, running silently in the back of his mind, and Tesla was watching for any opportunity to build it. That opportunity came in early 1883, in the form of a disaster.

* * *

The Continental Edison Company had installed a lighting plant at the new railway station in Strasbourg, in the German territory of Alsace. During the grand opening ceremony, in the presence of Kaiser Wilhelm I himself, a short circuit blew out a large section of the station wall. The German government refused to accept the plant. The French company was facing a serious financial loss.

Tesla spoke German. Tesla had experience troubleshooting power plants. Tesla was sent to Strasbourg to fix the mess.

He did fix it — the practical work, the correspondence with officials, the diplomatic negotiations between the French company and the German government occupied him day and night for months. But as soon as he could manage, he did something else. He found a mechanical shop across from the railway station, brought materials he'd carried from Paris, and built a motor.

Not just any motor. The motor. The AC induction motor he'd conceived in the Budapest park, refined in two months of ecstatic mental work, and carried in his head for over a year without being able to test it in the physical world.

It worked.

The rotor turned. No commutator. No brushes. No sparks. The rotating magnetic field swept around the stator and the rotor followed, exactly as Tesla had seen it in his mind on that February afternoon in the Városliget. The machine built in his imagination and the machine built in the Strasbourg shop produced the same result.

"In twenty years there has not been a single exception."

It's worth pausing here to register what this moment meant to Tesla. For over a year, the motor had existed only as a mental image — wonderfully sharp and clear, with the solidity of metal and stone, but still an image. Poeschl's voice was still in his ear: *Mr. Tesla may accomplish great things, but he certainly never will do this.* The six years of failure at Graz were still in his body. The breakdown was still in his nerves. And now, in a rented workshop in Strasbourg, the physical rotor was turning. The principle was real. The vision was correct. The professor was wrong.

Tesla tried to interest the people around him in what he'd done. The former mayor of Strasbourg, a man named Puskas (no relation to the Budapest Puskás), was sympathetic and put the invention before several wealthy citizens. None of them responded. A working AC induction motor was spinning in a shop in Alsace, and nobody with money could see what it was.

This would become the defining frustration of Tesla's life. The machine worked. The principle was sound. The implications were enormous. And the people with capital looked at it and shrugged.

* * *

By the spring of 1884, the Strasbourg plant was repaired, the diplomatic issues were settled, and Tesla returned to Paris expecting to be rewarded. One of the administrators had promised him a liberal compensation for the Strasbourg work, plus fair consideration for improvements he'd made to the company's dynamos.

What followed was a masterclass in corporate evasion. Tesla called on administrator A, who told him that administrator B had the authority. B thought only C could decide. C was quite sure that only A had the power to act. After several laps of what Tesla drily called this "circulus vicious," it dawned on him that his reward was, as he put it, "a castle in Spain."

He wasn't going to be paid. And his attempts to raise capital for the AC motor in Paris had gone nowhere.

There's a pattern here that's worth naming, because it'll repeat for the rest of Tesla's career. He does extraordinary work. The work is acknowledged. The reward is promised. And then the promise evaporates through a mechanism so polite and so bureaucratic that there's no one to confront, no villain to point to — just a system that absorbs his output and returns nothing. Edison would do it to him in New York. The investors would do it to him after the Tesla Electric Company. J.P. Morgan would do it to him at Wardenclyffe. The script barely changes.

Tesla wasn't built to navigate this. He understood electromagnetic fields with an intuition that bordered on clairvoyance. He couldn't read a room. The childlike quality that let him ignore Poeschl and persist with an "impossible" idea for six years was the same quality that left him standing in a Paris office, genuinely surprised that three administrators had conspired to avoid paying him.

Batchelor — Edison's man, the plant manager who'd watched Tesla work for two years — saw the situation clearly. He pressed Tesla to come to America. Edison's operation in New York needed someone who could redesign their DC dynamos, and Batchelor knew Tesla could do it. He wrote a letter of introduction.

The exact wording of that letter has become one of the great contested quotations in the history of technology. The version Tesla told, and that has been repeated ever since, is: "I know two great men and you are one of them; the other is this young man." Whether Batchelor actually wrote those words, or whether Tesla's memory polished them over the decades, is impossible to verify. The letter itself hasn't survived. What's certain is that Batchelor thought enough of Tesla to send him to Edison personally, and that Tesla, having exhausted every possibility in Europe, decided to go.

* * *

The journey to America was, fittingly, a small catastrophe.

Tesla sold his books and what few possessions he had, bought a train ticket and a steamer passage, and set off. At the station, he discovered that his luggage had been stolen, along with his ticket and most of his money. He barely talked his way onto the train. He barely talked his way onto the ship. He crossed the Atlantic in steerage with almost nothing. When he arrived in New York in June 1884, he had in his possession four cents, a sheaf of his own calculations, a few poems, a letter of introduction to the most famous inventor in the world, and the complete design for a machine that would make that inventor's entire empire obsolete.

He was twenty-eight years old. He'd never held a permanent position for more than a year. He didn't have a degree. He didn't have money. He didn't have connections beyond Batchelor's letter. He didn't speak the kind of English you learned in drawing rooms. And he was carrying, in his mind, an invention that virtually no one on the planet understood.

His first encounter in the New World was with a policeman. Tesla had been walking up Broadway from the docks, absorbing the scale of the city, when he passed a small machine shop where a man was trying to repair a piece of electrical equipment and cursing at it. Tesla stopped, offered to help, and fixed the machine in an hour. The owner gave him twenty dollars — five hundred times what he'd arrived with. America, at least, paid you when you solved the problem in front of you.

He found Edison the next day.

* * *

Thomas Alva Edison in 1884 was the most celebrated inventor in America and arguably the most famous private citizen in the world. He was forty-seven years old, enormously wealthy, and presiding over an electrical empire built entirely on direct current. His Pearl Street generating station in lower Manhattan, which had gone live in 1882, was the model for urban electrification. His name was on the dynamos, the distribution network, the light bulbs. He'd bet everything on DC, and in 1884 the bet appeared to be paying off.

Edison was also the anti-Tesla in almost every conceivable way. Where Tesla designed in his mind and built only when the design was perfect, Edison built first and iterated relentlessly — the famous ten thousand experiments to find the right light bulb filament. Where Tesla worked from principles, Edison worked from results. Where Tesla was tall, lean, fastidious, and spoke in the cadences of a European intellectual, Edison was stocky, unkempt, partially deaf, and proud of his lack of formal education. Where Tesla couldn't read a room, Edison could read a room, a boardroom, a newspaper editorial page, and the mood of the investing public with equal facility.

They met at Edison's headquarters on Fifth Avenue. Tesla presented Batchelor's letter. He described his work in Paris and Germany. And then, with the directness that was either his greatest strength or his greatest weakness, he told Edison about the rotating magnetic field and the polyphase AC system and how an enterprising man could make a great deal of money from it.

Edison's response, as Tesla recalled it, was immediate and absolute: "Hold up. Spare me that nonsense. It's dangerous. We're set up for direct current in America. People like it, and it's all I'll ever fool with."

And that was that. Tesla had walked into the office of the one man in America with the resources to develop polyphase AC power, had explained the principle clearly, and had been dismissed in a sentence. Edison didn't ask to see the diagrams. He didn't ask how the motor worked. He didn't want to know. AC was dangerous nonsense, and his mind was closed.

But Edison was overloaded with problems and short on engineers who could solve them, and he recognized immediately that Tesla was an exceptional engineer. He put him to work that same day, redesigning DC dynamos.

And so began one of the strangest working relationships in the history of technology: the man who had the future of electrical power complete in his head, employed by the man who'd built the present of electrical power and refused to hear a word about what came next. Tesla, the theorist who designed from principles. Edison, the empiricist who designed from trial and error. The two greatest electrical minds of their generation, in the same building, on the same problems, speaking what amounted to different languages.

It couldn't last. And it didn't.

* * *

But that confrontation belongs to the next chapter. For now, what matters is the arc of the journey itself: from the park in Budapest to the factory at Ivry to the workshop in Strasbourg to Edison's office on Fifth Avenue. At every stop, Tesla carried the same thing — a complete vision of a machine that would change the world — and at every stop, the world wasn't ready.

The resonance principle, which he'd applied so brilliantly to the problem of the commutator, didn't help him here. You can't find the natural frequency of a bureaucracy. You can't match the rhythm of venture capital or the mood of a railroad baron. Tesla's gift was for physical systems — systems that obeyed laws, that responded to precision, that rewarded anyone who understood their nature. Human systems don't work that way. They're noisy, irrational, and resistant to elegant solutions. And Tesla, the man who could feel whether an imaginary turbine was out of balance, couldn't feel the room going cold when he told Thomas Edison that his life's work was obsolete.

He'd need someone else for that. He'd need George Westinghouse. But first, he'd need to survive Edison.

CHAPTER FOUR

EDISON'S WORLD

New York, 1884–1887

To understand what Tesla walked into when he entered the Edison Machine Works on Goerck Street in lower Manhattan, you need to understand something about how Thomas Edison solved problems — because the contrast between the two men reveals exactly what made Tesla's mind extraordinary.

Edison's method was empirical. He stated it plainly and often, with a kind of pugnacious pride: "I have not failed. I've just found ten thousand ways that won't work." This wasn't self-deprecation. It was a philosophy. Edison believed that invention was a matter of persistence, variation, and volume. You tried things. You tried a lot of things. You recorded the results, discarded the failures, kept what worked, and tried again. The right answer was hiding somewhere in the space of all possible combinations, and if you searched long enough and hard enough, you'd find it.

It was a powerful method. It had produced the phonograph, the practical incandescent light bulb (after testing over three thousand filament materials), and the Pearl Street generating station that was electrifying lower Manhattan. Edison's instincts for what the public wanted were superb, his stamina was inhuman, and his ability to organise a team of machinists and engineers into a problem-solving factory was genuinely revolutionary. He'd invented the modern research laboratory. He deserved every bit of his fame.

But his understanding of electrical theory was, by Tesla's standards, crude. Edison worked from the outside in — from the observable behavior of the apparatus to the practical solution. He didn't particularly care *why* something worked, as long as it worked. He was suspicious of mathematics, contemptuous of theoretical physics, and openly hostile to alternating current, which he regarded as dangerous and unnecessary. His empire ran on direct current, and direct current was what he knew.

Tesla worked from the inside out. He started with the principle — the mathematical relationship, the physical law, the behavior of the field — and derived the machine from the principle. He didn't try things. He understood things, and then built what his understanding demanded. Where Edison searched through ten thousand possibilities, Tesla saw the answer before he'd touched a piece of wire.

These two approaches arrived in the same room in June 1884, and the encounter reveals everything about the nature of Tesla's gift.

* * *

Tesla's first real test at the Edison Machine Works came almost immediately.

The SS *Oregon* was an ocean liner — one of the first ships in the world fitted with Edison's electric lighting — and its dynamos had failed. The ship was in port and the lighting system was dead. Edison's own engineers, men who'd been building and maintaining these machines for years, couldn't locate the fault. Tesla was sent aboard.

He worked through the night. By dawn, both dynamos were running. Walking back to the shop in the early morning, he ran into Batchelor and Edison on the street. They made a crack about their "Parisian" being out all night. Tesla told them where he'd been. Edison turned to Batchelor and said: "This is a damned good man."

The episode is usually told as a story about Tesla's work ethic, and it is that. But the real point is what it demonstrates about how his mind worked. Edison's engineers approached the dead dynamos the Edison way — testing components, checking connections, working through possibilities one at a time. Tesla approached them the Tesla way. He saw the system as a whole — the magnetic circuits, the current paths, the thermal loads, the mechanical stresses — and located the failure by reasoning from the principle rather than by testing component after component.

This was the fire-pump moment from Gospić all over again. The experts were stumped. Tesla waded in and found the problem. Not because he knew more about those particular dynamos — he'd never seen them before — but because he understood electricity at a depth that made the specific machine almost irrelevant. If you understand how electromagnetic fields behave, you can diagnose any machine that uses them. The principle doesn't change. Only the hardware changes.

* * *

Edison put Tesla to work redesigning the company's DC dynamos. The existing machines were inefficient, expensive to maintain, and increasingly inadequate for the growing demand. Tesla threw himself into the task with his usual intensity — working from ten-thirty in the morning until five the next morning, seven days a week, no exceptions.

Edison, who'd seen hard workers before, was impressed. "I have had many hard-working assistants," he told Tesla, "but you take the cake."

Over the following months, Tesla designed twenty-four different types of standard machines with short cores and uniform patterns to replace Edison's old models. His notebook from this period shows wire diameters and weights computed to ten or more decimal places — a level of precision that had no practical meaning for the manufacturing process, but that was simply how Tesla's mind operated. Full volume, always.

What's really remarkable about the work isn't the hours or the precision. It's that Tesla was designing machines in a technology he didn't believe in — direct current, the very system his own invention was destined to replace — and he was doing it faster and better than the engineers who'd spent their careers on it. He could optimize a DC dynamo the way a concert pianist can play a simple tune: with total command and effortless clarity, because he could see the entire structure at a glance.

This is the point that matters. Tesla didn't just understand alternating current. He understood *electricity.* The rotating magnetic field, the polyphase system, the induction motor — these were his great inventions, but they emerged from something more fundamental: an intuitive, almost physical comprehension of how electromagnetic fields behave in space and time. He could feel the currents. He could see the fields. He could hold the entire system in his mind and watch it operate, and any deviation from the expected behavior told him immediately where the problem was and why.

Edison couldn't do this. Edison was a genius of a different kind — practical, commercial, relentless — but he solved electrical problems the way a locksmith picks a lock, trying each pin in turn until the mechanism gives. Tesla solved them the way a mathematician solves an equation: by seeing the structure. Both approaches work. But only one of them lets you conceive a complete power system in a flash of insight and know, before you've built a single prototype, that it'll work.

* * *

The great frustration of Tesla's time at Edison's shop wasn't the long hours or the modest pay. It was the silence.

Tesla had told Edison about the rotating magnetic field and the polyphase AC system on their very first meeting. Edison's response had been immediate and final: "Spare me that nonsense. It's dangerous. We're set up for direct current in America. People like it, and it's all I'll ever fool with."

And that was the end of the conversation. Not just for that day — for the duration of Tesla's employment. Edison didn't ask to see the diagrams. He didn't ask how the motor worked. He wasn't interested. Tesla later reflected on this with more wonder than bitterness: "This wonderful man, who had received no scientific training, yet had accomplished so much, filled me with amazement." It was a generous assessment from a man whose entire theoretical framework had just been waved away.

The irony is that Edison needed what Tesla was offering far more than he realized. Direct current couldn't be transmitted efficiently over long distances — the voltage losses in the wires were crippling, which meant you needed a generating station every mile or so. This wasn't a minor inconvenience. It was a fundamental physical limitation that would eventually make the DC model commercially unviable for anything beyond dense urban centers. Tesla's polyphase AC system solved this problem completely: transformers could step the voltage up for transmission and down for use, meaning a single power station could serve an entire region. The physics were clear, the mathematics were straightforward, and the solution was sitting in the same building as the man who most needed it.

But Edison had built an empire on DC, and empires don't pivot easily. This wasn't villainy. It was human nature. Edison was invested — financially, intellectually, emotionally — in the system he'd created, and the suggestion that it was fundamentally limited wasn't something he could hear, no matter who was saying it.

* * *

Tesla's time at the Edison Machine Works lasted about six months. He left in late 1884, partly over a dispute about a promised bonus that never materialized — a miscommunication about fifty thousand dollars that Tesla took at face value and the management may have intended as a joke. It was the sort of misunderstanding that follows people who are brilliant at systems and careless about contracts, and it would happen to Tesla more than once in his career.

What followed was a stretch of difficulty. Tesla briefly formed a company with investors who wanted him to design arc lamps rather than AC motors — a commercially sensible request that he fulfilled with his usual brilliance before the partnership dissolved. He spent a period doing manual labor, including digging ditches for Western Union, which he described with a single wry sentence: "My high education in various branches of science, mechanics and literature seemed to me like a mockery."

But Tesla wasn't a man who stayed down. He wasn't built for self-pity, and his confidence in the polyphase system never wavered. He kept talking about it to anyone who'd listen, and eventually the right people listened.

Through a series of introductions, Tesla met Alfred S. Brown, a superintendent at Western Union, and Charles F. Peck, a New Jersey lawyer with capital to invest. Both men were technically literate enough to follow Tesla's explanations and commercially astute enough to recognise what they were hearing. In April 1887, they formed the Tesla Electric Company, with Tesla receiving a third of the profits and his own laboratory at 89 Liberty Street in lower Manhattan.

For the first time since the park in Budapest, Tesla had what he needed: a place to build.

* * *

What happened next was one of the most concentrated bursts of invention in the history of technology.

Tesla had been carrying the polyphase system in his mind for five years. Every motor, every generator, every transformer had been designed, tested, refined, and perfected in his imagination. The machines had been running in his head since Budapest, accumulating improvements, shedding flaws, reaching a state of completeness that his visual system could certify with the same confidence a mathematician feels when a proof is sound. All he'd ever lacked was a workshop.

Now he had one, and the designs poured out of him.

In the space of a few months, Tesla built working models of virtually every type of polyphase AC motor and generator he'd conceived. Single-phase induction motors. Two-phase motors. Three-phase motors. Split-phase motors. Synchronous motors. Each one worked exactly as he'd designed it in his mind. Each one turned smoothly and silently, driven by the rotating magnetic field, without a commutator, without brushes, without sparks.

This is where the comparison with Edison becomes most instructive. Edison found the right light-bulb filament by testing thousands of candidates. It was heroic, it was effective, and it took years. Tesla built more than thirty patentable inventions in a matter of months, and not one of them required iteration. They worked the first time, because they'd already been working in his mind for half a decade. The mental turbine and the physical turbine produced the same results. Every time. Without exception.

By 1888, Tesla and his partners held more than thirty patents covering the entire polyphase AC system — generation, transmission, transformation, and motor design. It wasn't a single invention. It was a complete technology, from power station to factory floor, conceived as a unified whole and executed with a precision that left the electrical engineering community astonished.

This is the thing that gets lost in the popular telling, which tends to reduce Tesla's contribution to "he invented the AC motor." He didn't just invent a motor. He invented the *system*. The motor was the most visible part, but the genius was in the architecture — the recognition that alternating current, properly arranged into multiple phases, could do everything direct current could do and a hundred things it couldn't. Generation, transmission over vast distances with minimal loss, voltage transformation, smooth rotary power — all from a single unified principle. No one else saw it whole. Edison saw pieces. The European engineers saw pieces. Tesla saw the entire thing, complete, from the generator to the light switch, and he'd seen it since he was twenty-five years old, standing in a park in Budapest, watching the sun go down.

* * *

On the sixteenth of May, 1888, Tesla stood before the American Institute of Electrical Engineers and delivered a lecture entitled "A New System of Alternating Current Motors and Transformers."

The audience was the most technically sophisticated in the country. Tesla demonstrated working motors. He explained the rotating field. He showed how multiple phases of alternating current, each offset by a fixed angle, produced a smoothly rotating magnetic field that could drive a rotor without any mechanical switching. He showed that the system could work as a motor, a generator, or a transformer, and that all three functions derived from the same underlying physics.

The presentation was lucid, methodical, and quietly devastating. Tesla didn't argue. He didn't persuade. He *demonstrated.* The motors were spinning right in front of them. The mathematics were on the blackboard. The patents were filed. And everyone in that room understood that if the system worked as shown — and it manifestly did — then the entire future of electrical power had just shifted.

This was Tesla at his purest and his best: not negotiating, not politicking, not maneuvering for advantage, but simply showing what was true. Here is the principle. Here is the machine. Here is the proof. The clarity of the work spoke for itself, and for the first time, the right audience was listening.

* * *

George Westinghouse was listening. He wasn't in the room that night, but he read the reports within days. Westinghouse was the one man in America with the vision, the resources, and the appetite for risk to bet on alternating current. He'd already built the first AC power system near Boston. He'd been looking for an AC motor that worked. Tesla had just demonstrated that one existed, and that it was part of a complete system ready for the world.

Westinghouse sent his engineers to investigate. They confirmed everything. He came to New York, met Tesla, examined the patents, and made an offer: sixty thousand dollars for the patent rights — part cash, part stock — plus a royalty of two dollars and fifty cents per horsepower of electrical capacity sold.

Tesla accepted. He was thirty-two years old. Three years earlier he'd arrived in New York with four cents and a letter of introduction. Now his patents were the foundation of a new electrical order, backed by one of the most powerful industrialists in America.

The Westinghouse partnership wasn't just a business deal. It was the moment Tesla's vision found its medium. He'd had the principle since Budapest. He'd had the machines since Liberty Street. What he'd never had was the industrial capacity to build the system at scale —

the factories, the distribution networks, the political weight to go up against Edison's entrenched DC empire. Westinghouse provided all of that. He understood what Tesla had built, and he understood what it could become.

For the argument of this book, the Westinghouse deal marks a transition. Up to this point, the story of resonance has been the story of one mind — Tesla's — and its extraordinary ability to perceive the natural rhythms of electromagnetic systems. From this point forward, the story becomes larger. The principle that Tesla had carried in his head since a February sunset in the Városliget was about to be tested against the real world: against Edison's empire, against the skepticism of the establishment, against the physics of Niagara Falls.

The polyphase system was ready. The partnership was formed. What came next was the lecture circuit that would make Tesla famous, the war that would make him essential, and the waterfall that would prove him right.

CHAPTER FIVE

THE POLYPHASE SYSTEM

Pittsburgh, Telluride, and New York, 1888–1891

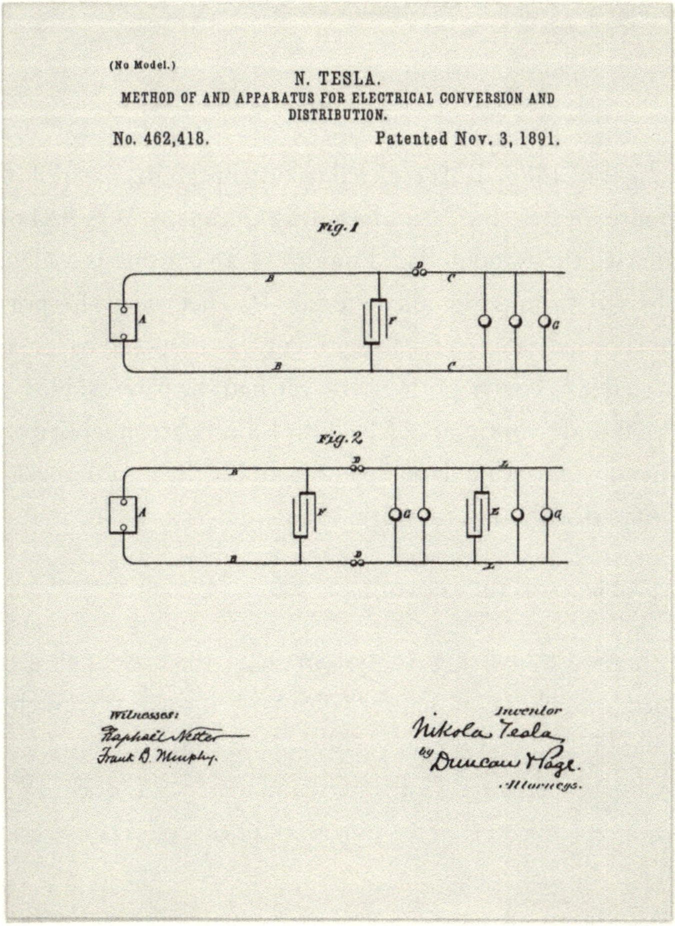

There's a moment in the life of every great idea when it has to stop being an idea and start being a product. The principle has to survive contact with reality — with manufacturing tolerances, with existing infrastructure, with engineers who didn't conceive it and don't think the way its inventor thinks. This is the moment that kills most inventions. It's the moment that nearly killed the polyphase system.

Not because the principle was wrong. The principle was perfect. But a perfect principle and a working power grid are separated by ten thousand practical decisions, and Tesla was about to discover that the gap between seeing the answer and building the answer at industrial scale was wider than he'd imagined.

* * *

In the summer of 1888, Tesla traveled to Pittsburgh to consult for the Westinghouse Electric and Manufacturing Company. Westinghouse was paying him two thousand dollars a month — an enormous fee, equivalent to more than fifty thousand dollars today — to help adapt his polyphase motors for commercial use. The assignment was supposed to last a year. Tesla expected it to be straightforward. He had the patents. He had the designs. He'd built working models in his Liberty Street laboratory. All that remained was to scale them up and connect them to the existing Westinghouse distribution network.

The problem was the network.

Westinghouse's existing AC system ran at 133 cycles per second. This frequency had been chosen for lighting — it was fast enough that incandescent bulbs didn't flicker visibly — and the entire installed base of Westinghouse generators, transformers, and meters was designed around it. Tesla's motors, however, were optimized for 60 cycles per second. At

133 cycles, the rotor speed was too high, the efficiency dropped, and the motors didn't perform as designed.

This was a resonance problem in the most literal sense. Tesla's motors were tuned to one frequency. The existing grid was tuned to another. And unlike a theoretical system running in his imagination, the physical system couldn't be wished into compliance. You couldn't just change the frequency of a thousand installed generators and every piece of equipment connected to them. Or could you?

The Westinghouse engineers, understandably, wanted to adapt the motor to the existing system. They'd been building 133-cycle equipment for years. Their meters worked at 133 cycles. Their transformers worked at 133 cycles. Changing the standard meant obsolescence — scrapping a generation of perfectly functional hardware and starting over.

Tesla's position was equally clear: the motor couldn't be adapted to 133 cycles without compromising its performance, and compromising the motor meant compromising the entire polyphase system. The frequency had to come down. Sixty cycles was the right number — it was the natural operating point of the motor, the sweet spot where the rotating field produced smooth, efficient torque without excessive rotor speed. The system had to be built around the physics, not the other way around.

It was, in miniature, the same argument he'd been making since Graz: don't fight the natural frequency. Match it.

* * *

The Pittsburgh experience was frustrating for everyone involved, and it's revealing about both the strengths and the limits of Tesla's particular kind of genius.

Tesla could see the complete system with absolute clarity. He knew what frequency the motors needed. He knew how the generators should be wound. He knew how the phases should be arranged. But translating that vision into manufacturing specifications, adapting it to existing tooling, negotiating compromises with engineers who had their own expertise and their own legitimate concerns — this wasn't his gift. Tesla was a principle man, not a factory man. He thought in fields and frequencies, not in production schedules and supply chains.

The Westinghouse engineers, for their part, were excellent practical men doing their best with a radically new technology. Charles F. Scott, who served as Tesla's assistant during this period and later became president of the AIEE, worked closely with him on motor design and made significant improvements to the slotted secondary winding. Benjamin Lamme, another Westinghouse engineer, would later use rigorous mathematical analysis to adapt Tesla's motors for industrial applications. These were serious engineers, and their contributions to making the polyphase system commercially viable were substantial.

But the frequency question was non-negotiable, and Tesla was right about it. Strenuous efforts to make his motor work at 133 cycles failed. As one contemporary account put it: "The little motor insisted on getting what it wanted, and the mountain came to Mahomet." Westinghouse eventually accepted Tesla's recommendation and adopted 60 cycles as the new standard.

It's a decision we still live with. Every electrical outlet in North America delivers power at 60 hertz — sixty cycles per second, the frequency Tesla specified in a Pittsburgh laboratory in 1889. Europe adopted 50 hertz, a close variant chosen for similar engineering reasons. The entire electrified world operates at frequencies that trace directly back to the operating characteristics of Tesla's induction motor. When you plug anything into a wall socket, you're connecting to a standard that a Serbian immigrant set because he understood the physics of rotating magnetic fields better than anyone alive.

* * *

The first real-world proof came in 1891, in one of the most unlikely places imaginable.

Telluride, Colorado, was a mining town perched at nearly nine thousand feet in the San Juan Mountains. The Gold King Mine needed power, and the mine was too remote and the terrain too rugged for a local coal-fired plant. But there was a river — the South Fork of the San Miguel — and where there's falling water, there's potential energy. The question was how to get the power from the river to the mine, several miles away.

Direct current couldn't do it. The transmission losses over that distance, at the voltages DC could manage, would have been crippling. But alternating current, stepped up to high voltage by a transformer, transmitted over the distance with minimal loss, and stepped back down at the other end? That was exactly the problem Tesla's polyphase system was designed to solve.

The Telluride installation, built using Westinghouse equipment based on Tesla's patents, was one of the first successful demonstrations of long-distance AC power transmission in the United States. A generator at the river produced alternating current, a transformer boosted the voltage, the power traveled several miles over wire to the mine, and another transformer brought it back down to a usable level. The mine ran. The lights came on. The principle worked, not in a laboratory and not in a lecture hall, but in the mountains of Colorado, in the dirt and the cold, powering real machinery for real work.

It was a small installation by later standards. But it proved the point that mattered: Tesla's system could do what Edison's couldn't. It could move power over distance. And if it could do it in Telluride, it could do it anywhere.

* * *

Tesla didn't stay in Pittsburgh for the full year. He was restless, impatient with the pace of industrial development, and his mind was already moving past the polyphase motor to something new.

This is one of the most striking features of Tesla's intellect, and it separates him from virtually every other inventor of his era. Most inventors, having produced a world-changing technology, would spend the rest of their careers refining it, commercializing it, defending the patents, and collecting the rewards. Tesla had the polyphase system — a technology that would eventually power civilization — and he was already bored with it. Not bored, exactly. Satisfied. The principle was proven. The motors worked. The frequency was set. The rest was engineering, and engineering, to Tesla, was the part that came after the interesting work was done.

What drew him back to New York was a question that had begun forming during the Pittsburgh work, a question that would define the next phase of his career: what happens when you push the frequency higher?

The polyphase system operated at 60 cycles per second. That was the right frequency for power transmission and motor operation. But what if you generated alternating current at hundreds of cycles per second? Thousands? Tens of thousands? What would happen to the behavior of the current, the magnetic fields, the relationship between electricity and light?

Tesla had an instinct — the same instinct that had told him the commutator was unnecessary, the same instinct that had told him the frequency mattered more than the force — that high-frequency currents would reveal something fundamental about the nature of electricity itself. The polyphase system harnessed alternating current at the frequency where it did useful mechanical work. But the spectrum didn't stop at 60 cycles. There were octaves above that, and Tesla wanted to know what was in them.

* * *

Back in New York, Tesla established a new laboratory — first at 175 Grand Street, then at 33–35 South Fifth Avenue — and began constructing high-frequency generators that could produce alternating current at frequencies far beyond anything the electrical industry had imagined.

The standard approach to generating AC used rotating machinery — a conductor spinning inside a magnetic field. This worked well up to a few hundred cycles per second, but mechanical generators couldn't spin fast enough to produce the frequencies Tesla was after. So he built a new kind of generator: machines with many poles arranged around the rotor, so that each revolution produced dozens or hundreds of cycles. By 1890, he was generating currents at frequencies up to twenty thousand cycles per second.

What he found astonished him and would astonish the world.

At high frequencies, alternating current began to behave in ways that defied common experience. It could pass through the human body without harm — a discovery Tesla demonstrated on himself, letting high-frequency currents flow through his hands while holding light bulbs that glowed brilliantly in his grip. It could illuminate vacuum tubes without any wire connection. It could produce light more efficiently than Edison's incandescent filaments. It could leap across gaps in circuits that would block ordinary current. It could produce heating effects, chemical effects, and mechanical effects that had never been observed before.

The electrical world had been operating in a narrow band of the frequency spectrum — DC at zero cycles, standard AC at 60 or 133 cycles — and Tesla was discovering that the spectrum extended upward into regions where electricity behaved like something entirely new. It was as if a musician who'd been playing in one octave suddenly discovered that the piano had eighty-eight keys.

* * *

It was in this period of high-frequency exploration that Tesla created the device that would become his most iconic invention and, for the purposes of this book, his most revealing: the Tesla coil.

The Tesla coil is a resonant transformer. That word — *resonant* — is the key to everything. An ordinary transformer transfers energy from one coil to another through electromagnetic induction, stepping the voltage up or down depending on the ratio of turns in the two coils. The Tesla coil does something more. It's tuned. The primary coil and the secondary coil are designed so that their natural electrical frequencies match, and when they match, energy transfers between them with extraordinary efficiency, building in amplitude with each oscillation until the voltage at the secondary reaches hundreds of thousands or even millions of volts.

It's the swing again. Push at the right moment — at the natural frequency of the system — and the amplitude grows with almost no effort. The Tesla coil is a machine that finds the electrical resonance of its own circuit and rides it, pumping energy from the primary to the secondary in precisely timed pulses until the voltage climbs to levels that would be impossible to achieve by brute force.

When the coil discharges, it produces artificial lightning — crackling, branching arcs of electricity that leap from the terminal into the surrounding air. The visual spectacle is extraordinary, and it's what most people think of when they think of Tesla. But the spectacle isn't the point. The point is the principle. The Tesla coil is a physical demonstration that resonance — the matching of a driving frequency to a natural frequency — can produce effects wildly out of proportion to the energy input. A few watts of power, properly timed, can generate voltages that would require enormous brute-force generators to achieve any other way.

Tesla understood this in his bones. The coil wasn't just an invention to him. It was confirmation. Everything he'd believed since childhood — that timing matters more than force, that matching the rhythm of a system unlocks its full potential, that the medium will do the work if you understand its nature — was embodied in this device. The polyphase motor had proven the principle at power frequencies. The Tesla coil proved it at radio frequencies. And Tesla was already thinking about what would happen if you scaled it up further — not to light a room, but to transmit power across a city, across a continent, across the world.

* * *

While Tesla was in Pittsburgh fighting about frequencies, something happened in Karlsruhe that would shape the next century of physics — and set the terms of a disagreement that runs through the rest of this story.

In 1887 and 1888, the German physicist Heinrich Hertz conducted a series of experiments that confirmed one of the great predictions of James Clerk Maxwell's electromagnetic theory: that oscillating electrical charges produce waves that radiate outward through space at the speed of light. Hertz built a simple spark-gap transmitter, generated electromagnetic oscillations, and detected them across his laboratory with a loop antenna. The waves behaved exactly as Maxwell's equations predicted. They radiated outward from the source in all directions. They reflected off surfaces. They refracted. They were transverse waves — oscillations perpendicular to the direction of travel — and they weakened with distance according to the inverse-square law, just as light does.

The discovery was electrifying. Hertz had demonstrated that light, radio waves, and all electromagnetic phenomena were fundamentally the same thing — waves radiating through space at a universal speed. The physics community embraced it immediately and completely. By the early 1890s, "Hertzian waves" was the standard term for radiated electromagnetic energy, and the entire framework for understanding how electrical signals travel was built on Hertz's model: energy radiates outward from a source, spreads in all directions, and weakens with distance. If you want to send a signal, you radiate as much energy as you can and hope enough of it reaches the receiver. Most of it won't. That's the nature of radiation.

Tesla had no quarrel with the physics. He'd produced Hertzian waves himself in his own laboratory. What he disputed — and would dispute for the rest of his life — was the engineering conclusion. Hertz had shown that electromagnetic energy radiates through space. The entire scientific establishment took this to mean that radiation was the mechanism for wireless transmission. Tesla drew the opposite conclusion. Radiation, he argued, was a loss mechanism. Energy that spreads in all directions is energy wasted. The vast majority never reaches the receiver. Any system built on Hertzian radiation would be inherently inefficient, fighting the inverse-square law at every step.

Tesla's alternative was conduction. Instead of radiating energy into space and hoping it arrives, drive it into a conducting medium — the earth — and guide it to where it's needed. A wave in a bounded medium doesn't spread in all directions. It follows the boundary. It doesn't weaken with the square of the distance. It behaves like sound in a pipe, not like light from a candle. The resonant transformer he'd just built — the Tesla coil — was, in this light, not merely a device for producing spectacular voltages. It was the first component of a system designed to couple energy into the earth itself and let the planet carry it.

This divergence — Hertz's radiation versus Tesla's conduction — is the fork in the road that defines the rest of Tesla's career. The world would follow Hertz. Radio, television, radar, cellular networks, satellite communication, Wi-Fi — all of it built on the principle of radiating energy into space from an antenna. Tesla would follow the other path: energy driven into the earth, guided through its body, received at a distance by a matched resonant receiver tuned to the planet's own frequency. Both paths began in the same physics. They led to entirely different civilizations. The one we built is Hertz's. The one Tesla designed was never tested.

* * *

By 1891, Tesla held more than forty American patents. The polyphase system was being commercialized by Westinghouse. The 60-cycle standard was established. The Telluride installation had demonstrated long-distance transmission. The Tesla coil had opened a new frontier of high-frequency electrical phenomena. And Tesla's name, which three years earlier had been known only to a small circle of electrical engineers, was beginning to reach a wider public.

He was thirty-five years old, at the peak of his powers, and about to do something no scientist or inventor had ever done before. He was about to take his high-frequency apparatus on tour — to New York, to London, to Philadelphia — and deliver a series of lectures that would make him the most famous electrical scientist in the world.

The lectures weren't just demonstrations. They were performances. Tesla, the awkward immigrant who couldn't negotiate a contract, turned out to be a mesmerizing public speaker who could hold a room full of scientists spellbound while passing a quarter of a million volts through his own body. The boy who'd stroked a cat in Smiljan and watched the sparks fly was about to show the world what electricity could do when someone truly understood it.

CHAPTER SIX

THE WIZARD OF PHYSICS

New York, London, Paris, and Philadelphia, 1891–1893

On the evening of May 20, 1891, Tesla stood before the American Institute of Electrical Engineers in the lecture hall at Columbia College, New York. He was thirty-four years old. He was not yet famous. The audience was composed of engineers and scientists who knew him, if at all, as the man behind the polyphase motor patents — an important figure in the industry, certainly, but not a household name. By the time he finished speaking, that would begin to change.

The lecture was titled *Experiments with Alternate Currents of Very High Frequency and Their Application to Methods of Artificial Illumination*. It was, on the surface, a technical presentation about generating alternating current at frequencies far above the commercial standard. But Tesla didn't just talk about high-frequency current. He demonstrated it. He showed the audience effects they'd never seen and couldn't explain. He lit vacuum tubes that glowed in his hands without any visible connection to a power source. He sent current through his own body and emerged unharmed. He produced brush discharges that resembled jets of luminous gas. He made wires glow, sparks dance, and the air itself seem alive with electricity.

The audience was astonished. The lecture was published in *The Electrical World* that July and immediately produced a sensation across Europe. Within months, Tesla was receiving urgent invitations from London, Paris, and Berlin. The continent's leading scientific societies wanted to see what the American journals were describing, and they wanted to see it in person.

* * *

To understand the impact of these lectures, you have to understand what the audience was seeing and why it mattered.

In 1891, most people's experience of electricity was limited to two things: the incandescent light bulb and the telegraph. Electricity was something that flowed through wires, heated filaments, and occasionally killed people. It was useful and it was dangerous, but it was fundamentally tame — a domesticated force, channeled through copper and controlled by switches.

Tesla showed them something wild.

He'd built generators capable of producing alternating current at frequencies of ten thousand cycles per second and higher — machines with hundreds of pole projections packed around the rotor, each revolution generating dozens of electrical cycles. He'd coupled these generators to his resonant transformers — the Tesla coils — which stepped the voltage up to hundreds of thousands of volts. And he'd discovered that at these extraordinary frequencies and potentials, electricity stopped behaving the way anyone expected.

At high frequencies, current would flow across the surface of conductors rather than through their centers. It would pass through the human body without causing muscular contraction or cardiac arrest — a phenomenon that seemed to contradict everything engineers knew about the lethality of electricity. It would illuminate gas-filled tubes with no wire connection, the tube glowing brilliantly when held near a charged conductor or even when simply placed in the electromagnetic field surrounding the coil. It would leap across gaps that would have blocked any ordinary current. It would produce light, heat, and mechanical rotation in ways that had no precedent in the existing literature.

Tesla didn't merely describe these effects. He performed them live, with himself as the primary experimental subject. He'd connect one hand to the terminal of his coil, hold an evacuated glass tube in the other, and the tube would blaze with light. Hundreds of thousands of volts were flowing through his body, and he stood there calmly, explaining the physics, while the audience watched in a state somewhere between scientific fascination and outright alarm.

* * *

The European tour began in London in February 1892. Tesla lectured first before the Institution of Electrical Engineers on the third of February, and the following evening, by special request, before the Royal Institution.

The Royal Institution lecture is one of those moments in the history of science that deserves more attention than it gets. The venue itself was extraordinary — the same lecture theatre where Michael Faraday had demonstrated electromagnetic induction six decades earlier, where

Humphry Davy had first produced the electric arc, where the greatest scientific minds of the nineteenth century had presented their work. The audience included Sir William Crookes, whose work on radiant matter had inspired Tesla as a student; Sir James Dewar, the institution's resident genius; and Lord Rayleigh, one of the foremost physicists of the age.

Tesla later recounted how the Royal Institution appearance came about. He'd intended to leave for Paris immediately after the IEE lecture, but Dewar wouldn't hear of it. The great Scotsman, as Tesla called him, physically pushed him into a chair, poured him a glass of whiskey, and announced that he was sitting in Faraday's chair and drinking Faraday's whiskey. Tesla, who was famously abstemious, apparently found both experiences irresistible.

The following evening, he gave the performance of his life.

He began with a tribute to Crookes, whose work he'd read in translation as a student at Graz and never forgotten. Then he proceeded to demonstrate effects that went far beyond anything even Crookes had imagined. He held evacuated tubes in his hands and lit them with the field from his coil. He showed phosphorescent substances glowing in tubes supplied by Crookes himself. He ran a motor on a single wire. He ran a motor with no wire connection at all. He produced lines of light seven inches long through open air. He brought a tiny platinum filament to incandescence inside a sealed tube and used it to spin a small fan — reproducing and extending Crookes's own famous experiment.

And through it all, he was calm, articulate, and unhurried. The man who couldn't navigate a contract in Pittsburgh, who'd been bewildered by the politics of the Westinghouse factory floor, turned out to possess an astonishing gift for scientific theatre. He knew how to build tension, how to sequence his demonstrations for maximum impact, how to explain what the audience was seeing in terms they could follow while making it clear that what they were seeing should be impossible.

When he finished, Lord Rayleigh addressed the audience. His words, Tesla later wrote, gave him the first real start in these endeavors. Dewar reportedly told the audience that Tesla had taken them into places in nature that only Crookes had previously explored. The *London Times* devoted a full column editorial to the lecture and published a detailed report of the demonstrations — an extraordinary level of coverage for a scientific event.

Tesla later recalled the evening with characteristic directness: the scientists, he said, simply didn't know where they were when they saw it.

* * *

What made the lectures so effective wasn't just the spectacle, though the spectacle was considerable. It was the fact that Tesla understood, at a level his audience didn't, exactly why these effects occurred.

The key was the skin effect. At high frequencies, alternating current tends to concentrate on the outer surface of a conductor. The higher the frequency, the thinner the layer of current flow. This is why Tesla could pass hundreds of thousands of volts through his body without harm: at the frequencies his coils produced, the current flowed across the surface of his skin rather than penetrating to his heart and muscles. He wasn't defying the laws of electricity. He was exploiting them.

He understood this because he understood resonance. The skin effect is itself a resonance phenomenon — at high frequencies, the electromagnetic field oscillates too rapidly for current to penetrate to the interior of the conductor, and the energy is effectively confined to the surface. Tesla didn't just know this as a theoretical fact. He knew it in his body, literally, because he'd tested it on himself long before he tested it on an audience. He knew precisely how far he could push the voltage, how high the frequency needed to be, and where the boundaries of safety lay.

This is a point worth dwelling on, because it says something important about the kind of scientist Tesla was. He wasn't reckless. He was intimate with his subject. He knew the behavior of high-frequency current the way a surfer knows waves — not from equations alone but from direct physical experience, from having felt it, having let it flow through him, having learned its character by letting it touch his skin. The showmanship was real, but it was built on a foundation of knowledge so deep that Tesla could stand at the edge of lethality and know, with absolute certainty, that he was safe.

* * *

From London, Tesla traveled to Paris, where he repeated his demonstrations before the International Association of Electrical Engineers and the French Society of Physicists. The reception was equally enthusiastic. Further lectures were planned for other European cities — Berlin, Vienna, Budapest.

Then a telegram arrived from his uncle Petar in Gospić. The message was brief and devastating: his mother was on her deathbed. He should come at once if he wished to see her alive.

Tesla canceled everything and rushed home. It was only the second time he'd returned to his homeland since leaving for America eight years earlier. He arrived in Gospić just hours before it was too late. Djuka, the woman who'd given him his visual imagination, who'd built her own tools and appliances without any formal education, who'd handed him a roll of bills during his gambling crisis and told him to go lose everything so he'd get it out of his system — she looked at her son and spoke the last words he'd ever hear from her.

You've arrived, Nidžo, my pride.

She died on Easter Saturday, April 4, 1892, at one o'clock in the morning. Six priests officiated at her burial in the Jasikovac cemetery in Divoselo, beside the grave of Tesla's father Milutin. Nikola raised matching white marble tombstones for both parents.

After her death, Tesla collapsed. He spent two to three weeks convalescing in Gospić and the nearby village of Tomingaj, his mother's birthplace. The pattern was familiar — the same kind of breakdown he'd suffered in Budapest a decade earlier, when exhaustion and emotional extremity overwhelmed a nervous system that was, in certain respects, more sensitive than most people's. The man who could pass a quarter of a million volts through his body was felled by grief.

He returned to New York in the summer of 1892, thinner, quieter, and more resolved than ever. He wrote to his uncle: *I am immeasurably sad, but console myself the best I can.* The European triumph and the personal loss seemed to crystallize something in him. From this point forward, Tesla would live with an even greater intensity of focus, an even deeper withdrawal from ordinary human entanglement, and an even more absolute commitment to his work. The lectures had shown the world what high-frequency electricity could do. Now he wanted to find out what it was for.

* * *

The final lecture of the series came in early 1893 — first at the Franklin Institute in Philadelphia on February 24, and then, at the request of the National Electric Light Association, repeated in St. Louis on March 1.

The St. Louis lecture was the most extraordinary of all, and not entirely by design. The original plan had been to limit attendance to association members, but the demand was overwhelming. The venue was expanded, and the audience swelled to more than five thousand people — far beyond anything Tesla had addressed before. This was no longer an audience of engineers and physicists. This was the public.

Tesla adjusted his presentation accordingly. The demonstrations were, if anything, more dramatic than what he'd shown in London. He showed new experiments, pushed the apparatus further, and produced effects so startling that at one point — as he later recounted with evident amusement — there was a stampede in the upper galleries. People rushed for the exits, convinced they were witnessing something diabolical. Tesla's own description of the scene was characteristically dry: *they thought it was some part of the devil's work.*

The Philadelphia and St. Louis lecture was titled *On Light and Other High Frequency Phenomena*, and it was the most ambitious of the three major lectures — not just in its demonstrations but in its theoretical scope. Where the Columbia lecture had focused on illumination and the London lecture on high-potential effects, the Philadelphia lecture reached toward the nature of light itself. Tesla argued that light and other electromagnetic phenomena were intimately connected to the behavior of the medium through which they traveled, and that by understanding high-frequency electrical oscillation, it might be possible to uncover the fundamental mechanism by which energy propagated through space.

It was here, for the first time in a public forum, that Tesla began to separate himself from the Hertzian consensus. Hertz had demonstrated that electromagnetic waves radiate outward from a source and weaken with distance. The physics community had taken this as the complete picture. Tesla was suggesting, carefully but unmistakably, that radiation was only part of the story — that the medium itself mattered, that the relationship between an oscillator and the space around it was richer and more complex than a simple outward broadcast of waves. He wasn't attacking Hertz. He was arguing that Hertz had described one phenomenon and the world had mistaken it for the only phenomenon.

He was reaching toward something enormous. Not just a better light bulb or a more efficient motor, but a unified understanding of electromagnetic energy — what it was, how it moved, and how it might be controlled at scales that dwarfed anything the electrical industry had imagined. The practical man of Pittsburgh, the showman of London, was revealing himself as something larger: a theorist with a vision that extended from the dynamo to the structure of the universe.

* * *

By the spring of 1893, Tesla was among the most famous scientists in the world. Not the most famous inventor — Edison still held that title in the public imagination — but something different, something that hadn't quite existed before: a physicist who was also a performer, a theorist who was also a showman, a man who could explain the deepest principles of nature while simultaneously demonstrating them in ways that bordered on the miraculous.

The lectures had accomplished something that no patent filing, no AIEE paper, no commercial contract could have achieved. They'd made Tesla's ideas visible. People who couldn't follow the mathematics of rotating magnetic fields could watch a glass tube light up in a man's hand and understand, viscerally, that something fundamental was happening. People who'd never heard of polyphase systems could see a motor spin without any wire connection and grasp, intuitively, that the old rules were being rewritten.

And the resonance thread ran through all of it. The Tesla coil that powered the demonstrations was a resonant device. The skin effect that kept Tesla alive was a frequency-dependent phenomenon. The phosphorescent tubes that glowed in his hands were responding to oscillations tuned to excite their contents. The motor that ran without wires was driven by an oscillating field matched to its natural frequency. Every demonstration, every effect, every moment of apparent magic came back to the same principle: find the natural frequency of the system, match it, and the system responds with an intensity that seems to come from nowhere.

Tesla wasn't performing tricks. He was performing physics. The lectures were the resonance principle made visible, made theatrical, made undeniable. And they set the stage for everything that came next.

Because Tesla, at the height of his fame, with the world's attention fixed on him, was already thinking past the lectures. He'd shown what high-frequency electricity could do in a room. He wanted to know what it could do across a continent. The boy who'd stood at the edge of Niagara Falls and imagined harnessing its power was now a man standing at the edge of something larger: the possibility of transmitting energy through the earth itself, without wires, using the planet's own resonant frequency as the carrier.

But first, there was a war to win.

CHAPTER SEVEN

THE WAR OF CURRENTS

New York, Chicago, and Niagara Falls, 1888–1896

While Tesla was building his high-frequency generators and preparing the lectures that would make him famous, a different kind of campaign was underway — one that had nothing to do with physics and everything to do with money, fear, and the lengths to which a great man will go when he feels his empire slipping away.

By the end of 1888, the numbers told the story. Edison's company was building and selling enough DC equipment to power forty-four thousand light bulbs per year. Westinghouse sold enough AC equipment to power forty-eight thousand bulbs in the month of October alone. Edison had built the first commercial power station in the world — Pearl Street, Manhattan, 1882 — but his direct current system could push electricity barely a mile from the generator before transmission losses became crippling. Westinghouse, armed with Tesla's polyphase patents and the transformer technology that made high-voltage long-distance transmission possible, was wiring cities that Edison's system couldn't reach.

Edison's response was not to build a better system. It was to make people afraid of the one that was beating him.

* * *

The campaign began on July 30, 1888, when an electrician named Harold P. Brown staged a public demonstration at Columbia College in New York. Brown had recently written an incendiary letter to the *New York Post* denouncing alternating current as a menace to public safety. Now he intended to prove it. He dragged a seventy-six-pound Newfoundland mix onto the stage, wired electrodes wrapped in wet cotton to its forelimbs, and subjected it to successive jolts of electricity — some DC, some AC — until the animal was convulsing against the cage. The audience was horrified. Several demanded he stop. Brown continued. The point was not science. The point was spectacle.

Brown was soon invited to use Edison's laboratory in West Orange, New Jersey. Someone posted a sign nearby offering twenty-five cents for stray dogs, and local men brought them in by the dozen — setters, terriers, Saint Bernards, bulldogs. Over the following months, Brown electrocuted dogs of every size and breed, then moved to larger animals. On December 5, 1888, before members of the press, the New York Medico-Legal Society, the chairman of the state's death penalty commission, and Edison himself, Brown electrocuted four calves and a twelve-hundred-pound horse using alternating current. The results were presented as scientific evidence that AC was uniquely lethal. In all, Edison's laboratory was the site of experiments that killed forty-four dogs, six calves, and two horses.

The extent of Edison's personal involvement was exposed in August 1889, when letters stolen from Brown's office and published in the *New York Sun* revealed that Brown had been receiving direction and payment from Edison's company. Edison had opened his laboratory to Brown and loaned him his top assistant. He'd written letters to city officials recommending Brown as an expert on AC safety. He'd imposed unusual secrecy conditions on the work — a striking departure from his normal policy of encouraging open exchange at his lab. Whether Edison conceived the campaign or merely enabled it, the documentary trail shows he was deeply embedded in its execution. Brown consistently denied any association with Edison. The stolen letters said otherwise.

The campaign escalated. Brown published a sixty-one-page pamphlet titled *The Comparative Danger to Life of the Alternating and Continuous Electrical Current*, detailing his animal experiments and claiming they proved AC was far deadlier than DC. The booklet — likely printed at Edison's expense — was sent to government officials, newspapers, and businessmen in every American city with a population above five thousand. Brown lobbied state officials to limit AC transmission voltages to three hundred volts, a level that would have made the technology commercially useless. And when New York State was developing the electric chair as a method of execution, Brown maneuvered to ensure it would be powered by alternating current — specifically, by Westinghouse generators that he acquired through intermediaries, since Westinghouse refused to sell equipment for that purpose.

The word "Westinghoused" was proposed in the press as a term for death by electrocution. The *New York Times* editorial page debated the terminology at length. The intent behind the coinage was transparent: to brand alternating current as the current of death, to make the public associate Westinghouse's name and Tesla's technology with execution and suffering.

* * *

On August 6, 1890, a convicted murderer named William Kemmler became the first person executed by electric chair, at Auburn Prison in New York. The procedure was catastrophic. The initial jolt of one thousand volts of alternating current didn't kill him. Kemmler was found to be still breathing, and the current had to be applied again — at higher voltage, for longer duration — while witnesses watched in horror. A reporter present described it as far worse than hanging. George Westinghouse, when informed of the outcome, offered a verdict that was characteristically blunt: *they would have done better using an axe.*

Edison had testified as an expert witness during the legal appeals preceding the execution, assuring the court that one thousand volts of alternating current would easily kill any man. His reputation as the

Wizard of Menlo Park carried enormous weight with the judges, and the appeals were denied. But the execution itself proved that the men administering it didn't understand the technology they were using. The voltage had been miscalculated. The electrodes were poorly placed. The entire apparatus had been designed by people who knew how to kill dogs but not how electrical current actually behaved at the scale of a human body.

It was, in a grim way, a precise illustration of the difference between the two camps. Edison's approach treated electricity as a brute force — you turned it up until it killed things, and that was the relevant fact. Tesla understood electricity as a phenomenon with specific behaviors at specific frequencies and voltages. He knew, from his own experiments on his own body, that current at the right frequency could pass through a man harmlessly, and that current at the wrong frequency and voltage could kill. The question was never whether electricity could be dangerous. Of course it could. The question was whether you understood it well enough to use it safely. And on that question, the war of currents was never really a contest.

* * *

Tesla himself took almost no part in the public battle. This is one of the most striking aspects of the war of currents: the man whose technology was at the center of the controversy stayed in his laboratory and worked.

He didn't write pamphlets. He didn't lobby politicians. He didn't stage counter-demonstrations. While Edison's camp was electrocuting animals and pushing for execution by Westinghouse generator, Tesla was building high-frequency alternators and discovering that electricity at the right frequency could light a room without wires. While Brown was mailing death statistics to every newspaper in the country, Tesla was preparing the lectures that would show a roomful of the world's leading physicists that the man who understood alternating current best could pass a quarter of a million volts through his own body and survive.

This wasn't passivity. It was a kind of strategy that only makes sense if you believe the technology will speak for itself when given the chance. Tesla believed that demonstration was more powerful than argument — that showing what AC could do was a better answer to fear than any amount of rhetoric. The record suggests he was right. His high-frequency lectures, delivered across America and Europe between 1891 and 1893, did more to dispel the fear of alternating current than any pamphlet or press release could have accomplished. When audiences saw Tesla standing unharmed in a field of artificial lightning, the argument that AC was inherently deadly became difficult to sustain with a straight face.

* * *

The decisive battle came in Chicago.

The World's Columbian Exposition of 1893, celebrating the four hundredth anniversary of Columbus's voyage, was to be the largest and most spectacular world's fair ever held. The organizers wanted the grounds illuminated by electric light — roughly a hundred thousand incandescent lamps, blazing after dark across six hundred acres of neoclassical pavilions and lagoons on the shore of Lake Michigan. The contract to light the fair would be the most visible demonstration of electrical technology the world had ever seen. Whoever won it would, in effect, light the future.

Edison General Electric — the company formed in 1892 from the merger of Edison's firm with Thomson-Houston — initially bid $1.72 million to power the fair with direct current, then slashed its bid to

$554,000 when a competitor emerged. Westinghouse, backed by a local Chicago company and using Tesla's polyphase system, came in at just under $399,000, underbidding GE by roughly seventy cents per lamp. The fair committee chose Westinghouse.

General Electric retaliated by banning Westinghouse from using the Edison incandescent light bulb, which was still under patent. It was a move designed to cripple the installation — the Edison bulb was the industry standard, and there was no obvious alternative. Westinghouse responded by engineering around the problem, designing a double-stopper bulb that circumvented Edison's patents. When Edison's company sued for infringement before the fair opened, they lost. The lights went up on schedule.

On May 1, 1893, President Grover Cleveland pressed a single button, and the White City blazed to life. The fair consumed three times more electricity than the entire rest of Chicago. Twelve thousand horsepower of polyphase AC generators — Tesla's design, built by Westinghouse — powered every light, every moving walkway, every elevator, every electric railway on the grounds. Over the course of the exposition, twenty-seven million people passed through the gates. They walked through a world made of light, and the light was alternating current.

Tesla came to the fair himself in August, attending the International Electrical Congress held at the Agriculture Building. In a specially constructed dark room within the Westinghouse exhibit, he performed his wireless lighting demonstrations for the general public — the same effects that had astonished the Royal Institution and panicked the upper galleries in St. Louis. He lit phosphorescent lamps without wires, produced crackling sheets of high-frequency discharge between insulated plates, and demonstrated the rotating magnetic field using his Egg of Columbus — a polished copper egg that stood upright and spun on a polyphase field, a physical metaphor for the motor that powered everything around them.

The fair was the turning point. Not because it revealed anything engineers didn't already know, but because it proved it to twenty-seven million witnesses. Even Lord Kelvin, the towering British physicist who had chaired the International Niagara Commission and had long been skeptical of alternating current, came to the exposition and left converted. After Chicago, the question of AC versus DC was functionally settled. Alternating current had won, and it had won the way Tesla always believed it would — not by argument but by demonstration, by the sheer undeniable fact of a hundred thousand lights burning safely across the White City, night after night, for six months.

* * *

There was one thing left to prove.

Since boyhood, Tesla had dreamed of harnessing Niagara Falls. He'd seen a photograph of the cataract as a child and told his uncle that one day he'd go to America and capture its power. The idea had seemed fantastical then. It didn't seem fantastical now.

The Niagara Falls Power Company, backed by investors including J.P. Morgan and W.K. Vanderbilt, had been studying the problem since the late 1880s. In 1890, they formed the International Niagara Commission under the chairmanship of Lord Kelvin, and solicited proposals from engineers and inventors around the world. The proposals ranged from compressed-air systems to rope-and-pulley arrangements to DC electrical transmission. Edison endorsed a direct current scheme. The commission rejected them all.

Then came Chicago. Kelvin, having watched the polyphase system power the exposition flawlessly for six months, reversed his long-standing position and threw his support behind the Westinghouse-Tesla approach. In late 1893, the commission awarded Westinghouse the contract to build the powerhouse and generators. General Electric received the contract for the transmission lines. It was a fitting division of labor: Tesla's technology would generate the power, and the infrastructure would carry it twenty-six miles to the city of Buffalo.

Construction took three years and was one of the most ambitious engineering projects of the nineteenth century. A hydraulic canal diverted water from the upper Niagara River into turbine chambers. A seven-thousand-foot brick-lined tailrace tunnel — requiring more than sixteen million bricks and costing the lives of twenty-eight workers over three years of construction — carried the spent water beneath the city to the lower river. Ten massive generators, each rated at five thousand horsepower, sat on vertical shafts connected to the turbines. The transformer system stepped the voltage up to twenty thousand volts for long-distance transmission, and back down to safe levels at the receiving end.

The first large generator was successfully tested on April 15, 1895. On August 26, 1895, the Adams Power Station went into operation, supplying electricity to local industries in Niagara Falls. Then, on November 16, 1896, the switch was thrown that sent alternating current surging twenty-six miles from the falls to the city of Buffalo. The streetcars ran. The lights came on. The *Niagara Falls Gazette* reported that the Niagara River was flowing uphill.

Tesla's childhood dream was fulfilled.

* * *

At the formal opening ceremony on January 12, 1897, Tesla spoke. His words captured something that went beyond the technical achievement, beyond the commercial victory, beyond even the personal vindication of a man who'd been told by Professor Poeschl in Graz that he would certainly never accomplish what he'd set out to do.

> *We have many a monument of past ages; we have the palaces and pyramids, the temples of the Greek and the cathedrals of Christendom. In them is exemplified the power of men, the greatness of nations, the love of art and religious devotion. But the monument at Niagara has something of its own, more in accord with our present thoughts and tendencies. It is a monument worthy of our scientific age, a true monument of enlightenment and of*

peace. It signifies the subjugation of natural forces to the service of man, the discontinuance of barbarous methods, the relieving of millions from want and suffering.

The subjugation of natural forces. But Tesla knew, better than anyone, that the word was slightly wrong. The Niagara generators didn't subjugate the falls. They didn't fight the water or overpower it. They worked with it. The river fell, the turbines turned, the generators spun, and alternating current flowed out at the frequency the system was tuned to produce. The power was already there — in the water, in the gravitational potential of a Great Lake emptying over a cliff. Tesla's contribution wasn't to create the power. It was to build a system that could receive it, that matched the rhythm of the falling water and converted it, through the rotating magnetic field, into a form that could travel twenty-six miles to light a city.

The swing again. The natural frequency of the system, identified and matched. The resonance principle, scaled from a child's toy to an industrial power plant, from a parlor demonstration to the infrastructure of civilization.

* * *

The Niagara plant generated fifty thousand horsepower — a tremendous output for its time, enough to power factories, streetcars, and the lighting of an entire city. Within a few years, the electricity was flowing as far as New York City, powering Broadway and the subways. Other plants were built on the Niagara and at sites across the country. By 1920, hydroelectric stations provided a quarter of all electricity in the United States.

The original Westinghouse generators ran for sixty-six years, from 1895 until the Adams Station was decommissioned in 1961, replaced by the Robert Moses Niagara Power Plant. One of the original generators now sits in the Smithsonian Institution. The transformer house, designed by McKim, Mead and White from locally quarried limestone, still stands — a National Historic Landmark since 1983, quietly deteriorating on the banks of the upper river.

Tesla was forty years old when the power reached Buffalo. He held more than a hundred patents. His technology lit cities, powered industry, and had defeated the most famous inventor in the world in the most public technological contest of the century. He was wealthy, celebrated, and at the absolute peak of his influence. He'd proven that alternating current worked, that the polyphase system worked, that the resonance principle could be applied at every scale from a laboratory coil to a continental power grid.

And he was about to do something that would seem, to the businessmen and financiers who'd backed him, completely insane. He was going to walk away from all of it — from the patents, from the royalties, from the guaranteed fortune that the polyphase system would generate for the rest of his life — and pour everything he had into an idea so ambitious that even his admirers would wonder if he'd lost his mind.

He was going to try to transmit power through the earth itself, without wires, using the planet as a conductor. And to do it, he was going to build the largest Tesla coil ever constructed, in the mountains of Colorado, and find out what happened when you tuned a resonant transformer to the electrical frequency of the planet.

CHAPTER EIGHT

THE MAGNIFYING TRANSMITTER

Colorado Springs, 1899–1900

On the afternoon of May 17, 1899, Nikola Tesla stepped off the Denver & Gulf train at Colorado Springs and into the thin mountain air, six thousand feet above sea level. He was forty-two years old, immaculately dressed as always, and obsessed with a single idea: that the planet Earth was a conductor, that it possessed a natural resonant frequency, and that a properly designed transmitter could set it ringing like a bell.

He'd come to Colorado to find out if he was right.

The move had been building for years. Tesla's laboratory on Houston Street in Manhattan was cramped, surrounded by buildings, and — after a fire in 1895 that destroyed much of his equipment and all his notes — unsuitable for the kind of work he now needed to do. What he planned required enormous power, enormous space, and air thin enough to ionize readily under high voltage. It required a place where artificial lightning wouldn't set neighboring buildings on fire. It required, in other words, somewhere that wasn't New York.

The invitation came from Leonard Curtis, a patent attorney who'd worked with Tesla and Westinghouse during the war of currents. Curtis held shares in the El Paso Power Company, the local electrical utility, and he offered Tesla everything an experimenter could want: free land on the open prairie east of the city, free electricity from the power company, and the influence to smooth any difficulties that might arise from a man building equipment capable of producing lightning bolts visible for miles. Tesla also secured thirty thousand dollars from Colonel John Jacob Astor and ten thousand from a dry-goods merchant named Crawford. He told his laboratory assistants to pack everything and ship it west.

A group of reporters was waiting at the Alta Vista Hotel on Cascade Avenue, where Tesla would stay for the duration. They wanted to know what he was planning. Tesla, who could be both compulsively forthcoming and infuriatingly evasive, gave them just enough: "I have come here to carry on a series of exhaustive experiments in regard to wireless telegraphy." Was he going to transmit signals from Pikes Peak? "No," Tesla replied. "I am here to work."

* * *

Construction began almost immediately. Tesla hired a local carpenter named Joseph Dozier and gave him specifications for what must have seemed like the strangest building in Colorado. The main structure was a barn-like shed, roughly fifty by sixty feet, braced on three sides, with a roof that could roll back to prevent it catching fire during experiments. Rising from the center of the roof was an eighty-foot wooden lattice tower, through which extended a metal mast reaching a total height of a hundred and forty-two feet above the ground, topped by a copper sphere thirty inches in diameter. Inside, the building was dominated by the largest Tesla coil ever constructed: a magnifying transmitter with a primary coil fifty-two feet in diameter, a secondary coil wound inside it, and an additional resonating coil — the "extra coil" — that functioned as a third stage of amplification, designed to drive voltages into the millions.

A high fence went up around the perimeter. Signs were posted on every post: **KEEP OUT — GREAT DANGER**. Tesla's assistant Fritz Lowenstein, who had a sense of humor, added another sign to the laboratory door, quoting Dante's *Inferno*: "Abandon hope, all ye who enter here."

The building went up in roughly two weeks. By mid-June, equipment was arriving from New York, including a Westinghouse transformer rated at fifty kilowatts. Tesla's assistants — Lowenstein and his trusted mechanic Kolman Czito — worked under his exacting direction to install and calibrate every component. Tesla kept a meticulous diary of every experiment, every measurement, every adjustment. The *Colorado Springs*

Notes, running to some five hundred pages and two hundred drawings, would become one of the most detailed laboratory records in the history of electrical science. It was never published in Tesla's lifetime.

* * *

To understand what Tesla built, and why it was different from anything that existed before, it helps to work through the apparatus from the ground up — literally.

A standard Tesla coil, the kind Tesla had been building and refining since the early 1890s, is a two-stage device. A primary coil, driven by an oscillating power source, induces energy into a secondary coil through electromagnetic coupling. Both coils are tuned to the same resonant frequency, so energy transfers efficiently from the first to the second, and because the secondary has many more turns of finer wire, the voltage is stepped up enormously — the same principle as any transformer, but operating at resonance, which means the energy doesn't just transfer once but builds over successive cycles. A well-tuned Tesla coil can produce voltages hundreds of times greater than the input. This is the device that powered every one of Tesla's high-frequency demonstrations, from the Columbia lecture to the Royal Institution to the pandemonium in St. Louis.

The magnifying transmitter was not a bigger Tesla coil. It was a fundamentally different architecture — a three-coil system that added a third resonant stage Tesla called the "extra coil." This is the key innovation, and it's what made the Colorado apparatus capable of effects that no two-coil system could achieve.

The primary coil sat beneath the floor of the laboratory — a massive ring fifty-one feet in diameter, wound from heavy cable capable of carrying enormous currents. The Westinghouse transformer drove this primary with fifty kilowatts of power at high voltage. The secondary coil was wound inside the primary, a cage-like structure of carefully spaced turns that rose to the height of the building. In a standard Tesla coil, the secondary would be the output. You'd draw your sparks from its top terminal and call it done.

Tesla didn't do that. Instead, the secondary was connected to the extra coil — a separate helical resonator, wound on its own form, tuned to the same frequency as the primary-secondary pair. The extra coil wasn't electromagnetically coupled to the other two; it was connected by wire, driven directly by the current from the secondary. But because it was tuned to resonance, it didn't just pass the energy through. It accumulated it. Each oscillation of the secondary fed energy into the extra coil, and the extra coil, vibrating at its natural frequency, stored that energy the way a bell stores the energy from repeated strikes. The voltage at the output of the extra coil rose with each cycle, limited only by the losses in the system — limited, that is, by the Q factor.

This is where the architecture becomes elegant. The bottom of the extra coil was connected to the ground — physically wired to a copper plate buried deep in the earth. The top was connected to the metal mast that rose a hundred and forty-two feet above the prairie, crowned by the copper sphere. The sphere and the mast served as an elevated capacitor: a large conductive surface raised high above the ground, which could accumulate enormous electrical charge. The ground connection at the bottom and the elevated terminal at the top created a voltage gradient — millions of volts of difference between the earth and the sky, oscillating at the transmitter's resonant frequency.

This is the point the popular accounts miss entirely. The spectacular lightning bolts that erupted from the copper sphere were a side effect

— a dramatic one, but a side effect nonetheless. The real output of the magnifying transmitter went downward, not upward. The oscillating voltage drove enormous currents into the earth through the ground connection, and those currents propagated outward through the planet as electromagnetic waves. The mast and sphere were not the antenna in the radio sense of broadcasting energy into the air. They were the top plate of a capacitor whose bottom plate was the earth itself. The system was designed to pump energy into the ground, to set the planet vibrating at the transmitter's frequency — to excite, in Tesla's own language, the stationary waves he'd detected.

Viewed this way, the magnifying transmitter was a resonant pump. The Westinghouse transformer provided raw power. The primary coil converted it to high-frequency oscillations. The secondary stepped up the voltage. The extra coil, operating at resonance, amplified it further through Q-factor accumulation. And the ground connection injected the result into the earth at a frequency matched to the planet's own resonant characteristics. Each stage was a resonant system coupled to the next, and every coupling was tuned so that energy flowed in one direction: from the source, through three stages of amplification, and into the planet.

There was one more design choice that distinguished the magnifying transmitter from anything else being built in 1899, and it was deliberate. Tesla drove his primary circuit not with smooth, continuous alternating current, but with abrupt, impulsive discharges from a spark gap — a technique he called disruptive discharge. A capacitor bank charged to high voltage and then released its energy in a single violent pulse across a spark gap, slamming the primary coil with a burst of current that rose almost instantaneously and decayed in a rapid oscillation. The process repeated many times per second, each discharge ringing the primary like a hammer striking a bell.

This was not a workaround for the lack of better oscillators. Tesla had built smooth mechanical alternators capable of producing continuous high-frequency current — the same machines that had powered his lectures. He chose disruptive discharge because it did something continuous oscillation could not. An abrupt impulse excites a resonant system across a broad spectrum of frequencies simultaneously, driving energy into the natural mode with maximum efficiency. Think of the difference between gently rocking a bell and striking it sharply. The strike transfers energy into the bell's resonant mode in a single instant; the bell then rings at its own frequency regardless of how it was hit. Tesla understood that the same principle applied to his transmitter and, by extension, to the Earth. An impulsive discharge coupled energy into the ground more efficiently than a smooth sinusoidal wave, because the impulse contained the resonant frequency of the target system and the target system selected it out. The medium did the tuning.

This choice had a further consequence that Tesla considered fundamental. A continuous sinusoidal oscillation radiates energy into space as electromagnetic waves — Hertzian waves, in the terminology of the day. That radiated energy is lost. It propagates outward in all directions and dissipates. Tesla didn't want to radiate energy into space. He wanted to drive it into the ground. The sharp, impulsive character of the disruptive discharge produced a different kind of wave — one that coupled into the conductive body of the Earth rather than radiating away from it. The distinction between radiation and conduction was not incidental to Tesla's design. It was the whole point.

The whole apparatus was, in effect, a quarter-wave resonator. The electrical length from the ground connection to the top of the elevated terminal was designed to be one quarter of the wavelength at the operating frequency. This is the same geometry that makes a quarter-wave antenna efficient — it creates a natural impedance match between the oscillator and the medium it's driving. In Tesla's case, the medium was the Earth. The tower didn't radiate energy into space like a radio antenna. It drove energy into the ground like a tuning fork pressed against a table, coupling its vibrations into the larger body through direct contact.

Tesla expressed the central idea with characteristic clarity in a later interview: "The idea presented itself to me that it might be possible, under observance of proper conditions of resonance, to transmit electric energy through the Earth, thus dispensing with all artificial conductors." The magnifying transmitter was his instrument for testing that idea. Everything about its design — the three-coil architecture, the resonant tuning, the ground coupling, the elevated terminal — served a single purpose: to deliver maximum energy into the planet at the planet's natural frequency, and to see what happened.

* * *

The first experiments were conducted on June 15, 1899. Tesla recorded the initial spark length at five inches — short, but thick and noisy. He had a lot of calibrating to do. The ground connection alone required digging a hole twelve feet deep, laying a copper plate twenty inches square on the bottom, packing it with coke, and running water constantly to reduce the resistance of the dry, rocky soil. Even then, the connection was poor. It took days of adjustment before the resistance dropped to fourteen ohms. Tesla noted that the rocky formation and dryness explained why lightning strikes in Colorado were so often destructive — the earth itself was a poor local conductor, which made the standing-wave effects he was looking for all the more dramatic when they appeared.

* * *

What Tesla was looking for, and what he found, was the most important discovery of his time in Colorado and arguably the most consequential experimental result of his life.

He'd arranged his receiving instruments to detect electrical disturbances in the earth. This was preparatory work — he wanted to understand the planet's electrical behavior before attempting to drive it. What he found astonished him. "The earth was found to be, literally, alive with electrical vibrations," he wrote. Colorado's violent summer thunderstorms, which rolled across the plains with tremendous regularity, provided natural laboratories of electrical force. Tesla could sit in his shed and watch the instruments respond to lightning strikes miles away, the signals strengthening and weakening in patterns that told him something extraordinary.

The decisive observation came on July 3, 1899 — a date Tesla said he would never forget. A massive storm gathered in the mountains to the west and, as evening came, moved out across the plains with tremendous velocity. Heavy, persistent lightning arcs formed at nearly regular time intervals. Tesla, his instruments already calibrated from weeks of observation, tracked the electrical effects as the storm retreated. The signals grew fainter with distance, as expected — and then, when the storm had passed roughly three hundred kilometers away, the signals stopped.

But they didn't stay stopped. In regularly recurring intervals, the same pattern of disturbance repeated. Fainter, then stronger, then fainter again. The effect continued long after the storm itself had passed beyond any reasonable range. Tesla's assistant Fritz Lowenstein independently confirmed the observations. And in the days that followed, additional storms provided further opportunities that brought the phenomenon out with unmistakable clarity.

Tesla had discovered stationary waves in the earth.

To understand what this means — and why it mattered so enormously to Tesla — it helps to picture an experiment that any physics student would have recognized in 1899, and that remains one of the most beautiful demonstrations in science. Take a flat metal plate, clamp it at its center, scatter fine sand across the surface, and draw a violin bow along one edge. The plate vibrates, and the sand dances — but not randomly. It collects into precise geometric patterns: lines, curves, nested shapes of startling elegance. These are called Chladni figures, after the German physicist Ernst Chladni, who first displayed them in the late eighteenth century. Napoleon was so impressed by a demonstration that he offered a prize for their mathematical explanation.

What the sand reveals is the plate's standing wave pattern. When the bow excites the plate at one of its natural frequencies, waves propagate outward, reflect from the edges, and interfere with the incoming waves to produce a stable pattern of vibration. Some regions of the plate oscillate vigorously — these are the antinodes. Other regions remain perfectly still — these are the nodes, and this is where the sand collects, because it's shaken off the moving parts and gathers where nothing moves. Change the frequency of the bow, and the pattern changes: different frequency, different geometry, different arrangement of nodes and antinodes. But at any given frequency, the pattern is fixed and repeatable. The plate rings at the frequencies its geometry permits, and at no others.

Now here is the point that mattered to Tesla. A traveling wave — one that propagates outward and never returns — carries energy away. It dissipates. It weakens with distance, and the energy it carries is lost to the medium. But a standing wave doesn't travel. It stays. The energy in a standing wave isn't being carried away and dissipated; it's being stored, oscillating back and forth between two forms — kinetic and potential in a vibrating plate, electric and magnetic in an electromagnetic system — without leaving the structure. A standing wave is, in effect, a pattern of energy that the system holds rather than loses. This is why a bell keeps ringing after you strike it, why a tuning fork sustains its note, and why a wine glass can be shattered by a singer holding exactly the right pitch: the energy doesn't leak away. It accumulates.

What Tesla's instruments told him, that July evening, was that lightning strikes were doing to the Earth what a violin bow does to a Chladni plate. The electromagnetic pulses produced by the storm propagated through the planet, reflected from the far side, and interfered with the outgoing waves to produce stable patterns of nodes and antinodes — regions of maximum and minimum electrical activity, fixed in space relative to the source. As the storm moved away, his receiving circuit passed successively through these nodes and antinodes, which explained the rhythmic strengthening and weakening of the signal. The Earth was ringing. Not chaotically, not briefly, but in a sustained, structured, standing-wave pattern determined by the planet's size and electrical properties.

The implication was staggering. In Tesla's own words, written for the *Electrical World and Engineer* in 1904: "No doubt whatever remained: I was observing stationary waves. As the source of disturbances moved away the receiving circuit came successively upon their nodes and loops. Impossible as it seemed, this planet, despite its vast extent, behaved like a conductor of limited dimensions."

A conductor of limited dimensions. Not a vast reservoir that absorbs disturbances and goes quiet, as most physicists assumed — not an ocean that swallows every stone thrown into it. A conductor with a finite size, a definable geometry, and therefore a set of natural frequencies at which it would ring, just as the Chladni plate rings at the frequencies its shape and material permit. And if the Earth could sustain standing waves, then energy injected at the right frequency wouldn't dissipate. It wouldn't weaken with distance the way a traveling wave weakens. It would be held in the structure of the planet itself, oscillating between nodes and antinodes with minimal loss — available to be tapped at any point along the pattern by a receiver tuned to the same frequency.

This was the insight that drove everything Tesla did from this point forward. A transmitter tuned to the Earth's resonant frequency could pump energy into the planet the way a singer's sustained note pumps energy into a wine glass. Small, repeated impulses at precisely the right frequency would build cumulative amplitude — not because the transmitter was powerful, but because the Earth itself would store and amplify the energy through resonance. The child pushing the swing, again. But this time the swing was the planet.

But Tesla knew that the mere existence of standing waves wasn't enough. A drum sustains standing waves, too, and a drum goes quiet in seconds. What mattered was not just whether the Earth could ring, but how long it kept ringing — how many times the energy would circulate before friction, radiation, and other losses bled it away. In engineering, this property has a name. It's called the Q factor.

The Q factor — the quality factor — of a resonant system is the ratio of the energy stored in the system to the energy lost per cycle of oscillation. A system with a Q of ten stores ten cycles' worth of energy before the losses catch up. A system with a Q of a thousand stores a thousand cycles' worth. The higher the Q, the longer the system rings, and — critically — the more energy accumulates before the input and the losses reach equilibrium. A cheap tuning fork might have a Q of a few hundred; it hums for a moment and dies. A quartz crystal has a Q in the tens of thousands; tap it once and it vibrates for minutes. The distinction isn't academic. It's the difference between a system that wastes most of the energy you feed into it and a system that hoards it.

This is why Q factor was the question that mattered most to Tesla at Colorado Springs, and why his stationary wave observations were so electrifying. If the Earth's electromagnetic Q factor was low — if the planet absorbed and dissipated electrical energy quickly, the way a wet towel absorbs sound — then wireless power transmission would be hopelessly inefficient. You'd pump energy in and it would vanish into heat and radiation before a distant receiver could capture any meaningful fraction of it. But if the Q factor was high — if the planet was more like a bell than a drum — then energy injected at the resonant frequency would circulate and accumulate, building in amplitude with each cycle of the transmitter. A modest transmitter, running continuously, could establish a powerful standing wave pattern across the entire globe, and receivers anywhere on the surface could tap into it the way a radio picks up a broadcast.

Tesla's Colorado measurements suggested the Q was high. The stationary waves he detected persisted long after the lightning that produced them had dissipated. The planet was ringing, and it was ringing clearly — not the muddy, quickly-damped response of a lossy system, but the sharp, sustained resonance of a system that held onto its energy. Later analysis by researchers James and Kenneth Corum, working from Tesla's diary data, estimated that the Earth-ionosphere cavity has a Q factor in the range of several hundred to over a thousand at the frequencies Tesla was working with. This is not the Q of a quartz crystal, but it's enormously higher than what skeptics assumed. It means the Earth doesn't just sustain standing waves — it sustains them efficiently. Energy put into the system at resonance circulates hundreds of times before dissipating.

This is the quantitative heart of Tesla's wireless power concept, and it's worth stating plainly. The feasibility of the entire scheme — Wardenclyffe, the World Wireless System, everything that came after Colorado — depends not on any exotic physics but on a single measurable number: the Q factor of the Earth at the relevant frequency. If it's high enough, the system works in principle. If it's too low, it doesn't. Tesla believed his measurements showed it was high enough. The question of whether a full-scale system could achieve practical efficiency at planetary distances — whether the engineering losses of coupling, transmission, and reception would eat the theoretical advantage — was never tested, because the full-scale system was never built. But the underlying physics, the Q factor that makes the whole idea conceivable, was measured at Colorado Springs. That measurement is what Tesla carried back to New York.

* * *

With the discovery of stationary waves confirmed, Tesla turned to the second purpose of the Colorado laboratory: to find out what happened when you drove a magnifying transmitter at full power and pumped electrical energy directly into the ground.

The experiment took careful preparation. Every piece of equipment was checked and rechecked. Tesla had calculated the dimensions of his apparatus with obsessive precision — the coil's geometry, the capacitance, the inductance, the resonant frequency of every component in the chain. When it was tuned correctly, energy would cascade through the three stages of the transmitter with terrific efficiency, each stage amplifying the oscillation until the voltage at the top of the mast reached millions of volts. Tesla later described the fifty-two-foot coil in operation: the streamers passed from top to bottom and covered perhaps fifteen hundred square feet of surface, coming within six or seven inches of the walls of the building. He'd calculated the dimensions as closely as possible to save money. The margin of safety was literal inches.

On the evening of the experiment, Tesla put on a Prince Albert coat and a black derby hat, as though dressing for the occasion were part of the protocol. He positioned himself outside the building where he could observe the effects. Czito stood inside at the heavy switch. Tesla gave the first instruction: close the switch for one second only. Czito threw it. The secondary coil erupted in sparkle and crack, and an eerie blue corona formed in the air around the apparatus. One second. The switch opened. Everything was working.

Tesla gave the second instruction: close the switch and hold it until told to stop.

What followed was the most dramatic electrical display ever produced by human hands. Huge arcs of blue electricity snaked up and down the center coil. Bolts of artificial lightning — some accounts say a hundred feet, others a hundred and thirty-five feet in length — exploded from the copper sphere at the top of the mast and cracked across the Colorado sky, producing thunder that was heard fifteen miles away. The earth shook with the intensity of the apparatus operating in undamped mode. A blue corona enveloped the entire building. Horses grazing half a mile away reared on their hind legs and bolted. Passers-by saw sparks leaping from the ground to their feet. Butterflies near the laboratory flew erratically, their wings bridged by tiny discharges. The lightning arresters on nearby buildings crackled with sympathetic energy. People as far as ten miles away later reported seeing sparks.

And then, abruptly, it all stopped. The roar ceased. The lightning vanished. Everything went dark.

"Czito!" Tesla shouted. "I am not ready for you to cut off the power!"

"I didn't!" Czito shouted back. "There is no power."

Tesla's magnifying transmitter had drawn so much current that it burned out the generator at the El Paso Power Company's station, six miles away. The dynamo had erupted in flames. The entire city of Colorado Springs lost power. Tesla immediately telephoned the power company and demanded they repair the generator. The company's manager was, understandably, livid. The free electricity arrangement was over. From that point forward, Tesla paid for his power — or at least was supposed to.

* * *

The generator burnout is the incident most often repeated in popular accounts of Colorado Springs, and it's usually told as a moment of mad-scientist excess — Tesla overreaching, blowing things up, unable to control his own creation. The truth is more interesting and more revealing.

Tesla didn't lose control. The magnifying transmitter did exactly what it was designed to do. It amplified electrical oscillations through three resonant stages until the energy at the output dwarfed the energy at the input — not by creating energy from nothing, but by accumulating it through resonance, the way a tuning fork builds amplitude from a sustained tone. The problem wasn't the transmitter. The problem was that the power company's generator wasn't built to supply the kind of current the transmitter demanded at peak resonance. Tesla had found the capacity limit of the local grid, and he'd found it by exceeding it. This is not recklessness. This is what rigorous engineering looks like when you are operating at a scale nobody has operated at before: you discover the limits of the surrounding infrastructure by being the first thing that exceeds them.

What the experiment proved was that the magnifying transmitter's three-stage resonant architecture worked at enormous scale. The hundred-foot lightning bolts weren't brute force. They were the product of precision tuning — energy cascading through three resonant stages, each one amplifying the oscillation, with the Q factor of each stage multiplying the cumulative effect. The same principle that had powered every Tesla coil since Houston Street, but driven to a level that turned night into day over the Colorado prairie. And the real output — the current driven into the ground — was even more impressive than the lightning, because that was the energy doing what it was designed to do: entering the planet.

Tesla also demonstrated, during his nine months in Colorado, something that conventional electrical theory said shouldn't work. He transmitted energy through the earth without wires.

He drove his magnifying transmitter and lit incandescent lamps stuck into the ground at distances of up to twenty-six miles from the laboratory, powered solely by currents flowing through the earth from the transmitter's ground connection. No overhead wires. No return circuit. No conductor of any kind linking the transmitter to the lamps. The most complex demonstration involved two hundred fifty-watt bulbs — ten thousand watts of total load — illuminated by energy that had traveled through the body of the planet. The lamps had only a single connection: a wire running from their terminals into the ground. Tesla's transmitter had only a single connection to the earth at its end. The circuit, such as it was, ran through twenty-six miles of rock and soil.

This was not merely a spectacle. It was a direct experimental test of Tesla's core theoretical position — a position that put him in sharp opposition to the prevailing understanding of how electromagnetic energy propagates.

By 1899, the fork in the road described earlier had hardened into orthodoxy. The scientific establishment had fully embraced Hertz's framework: electromagnetic waves radiate outward from a source as transverse oscillations, spreading in all directions and diminishing with distance according to the inverse-square law. Hertzian waves were the basis of Marconi's wireless telegraphy, which transmitted signals by radiating energy into the atmosphere from elevated antennas. The more energy you radiated, the farther the signal reached — but it weakened rapidly with distance, because radiation spreads in all directions and most of it is lost.

Tesla's position, as we've seen, was that this was the wrong mechanism entirely. He didn't dispute that Hertzian waves existed. What he disputed was that they were useful for practical transmission at any meaningful scale. Radiation dissipates. It spreads. It weakens with the square of the distance. Any system based on it is inherently wasteful, because the vast majority of the radiated energy never reaches the receiver. Tesla argued that effective wireless transmission required not radiation but conduction — energy driven into the earth and guided through it as current, the way water flows through a pipe rather than evaporating into the air. The Earth, in Tesla's analysis, was not an obstacle to wireless transmission. It was the medium. The Colorado experiments were his opportunity to prove it.

The Colorado experiments bore this out. The twenty-six-mile demonstration didn't work by radiation. There was no elevated antenna broadcasting energy into the atmosphere. The transmitter drove current into the ground through a buried plate, and the receivers extracted current from the ground through buried connections. The energy traveled through the earth. It arrived at the receivers with enough power to light two hundred bulbs. And it did so without a return wire — because in a resonant conduction system, the return path is the earth itself, just as the return path for a wave in a bell is the body of the bell. You don't need a separate conductor to bring the energy back. The standing wave pattern ensures that energy is available at the antinodes, and a tuned receiver placed at an antinode extracts it directly.

Tesla would later write, in "The True Wireless" in 1919, that Marconi's transatlantic transmissions actually worked by the same mechanism — ground conduction — whether Marconi knew it or not. The elevated antennas and the Hertzian interpretation were, in Tesla's view, a misreading of what was actually happening. The energy that reached Marconi's receivers had traveled through the earth and the surface waveguide, not through empty space. Tesla's argument was not merely theoretical. He had built a system that operated explicitly on the conduction principle, and it worked. The lamps lit. The energy was transmitted. Twenty-six miles, through rock and soil, without a single wire.

And in a way, every person who walked near the laboratory during those months was an unwitting proof of the same principle. The reports of sparks leaping from the ground to people's shoes, of horses rearing as tiny shocks prickled through their hooves, of butterflies trailing faint discharges from their wings — these weren't side effects of energy radiating through the air. If the transmitter had been broadcasting Hertzian waves from the mast, you would feel nothing underfoot. The sparks came from below, from the earth, because that is where the energy was. Every passerby was a miniature version of the apparatus itself: the ground beneath their feet was the charged conductor, their body the elevated terminal, and the voltage gradient between the two produced the same effect as the gap between Tesla's buried copper plate and the sphere atop his mast — the same physics, the same mechanism, measured in shoe leather instead of a hundred and forty-two feet of tower.

* * *

And then there were the signals.

One night during the summer of 1899, while monitoring his sensitive receiving equipment for electrical disturbances, Tesla detected something he couldn't explain. His instruments registered a series of rhythmic pulses — regular, repeating, and structured in a way that seemed to follow a counting pattern: one pulse, then two, then three. The signals were faint. They were unlike anything produced by thunderstorms or by his own apparatus. And they recurred.

Tesla first mentioned the signals in a letter to the journalist Julian Hawthorne at the *Philadelphia North American* on December 8, 1899. In December 1900, in a letter to the Red Cross Society about possible discoveries in the new century, he referred to messages "from another world" that read "1 ... 2 ... 3." In a February 1901 article for *Collier's Weekly* titled "Talking with Planets," he expanded on what he'd observed, noting that it had not been immediately apparent to him that he was hearing intelligently controlled signals, and that their source could be Mars, Venus, or other planets.

The press treated this as sensational evidence that Tesla was hearing Martians. Tesla himself was more measured than the headlines suggested — he described the signals carefully, acknowledged uncertainty about their origin, and framed the question as one that deserved investigation rather than premature conclusion. But the damage to his reputation was immediate and lasting. The man who had lit the White City and conquered Niagara was now, in the public mind, the man who talked to Mars.

What did Tesla actually detect? Several explanations have been proposed. He may have intercepted signals from Marconi's wireless telegraphy experiments in Europe — Marconi was actively transmitting in the summer of 1899, and the letter S in Morse code (three dots) would match the "1 ... 2 ... 3" pattern. A more intriguing hypothesis, advanced by researchers replicating Tesla's experiments a century later, suggests that what he received were natural radio emissions produced by the moon Io passing through Jupiter's magnetic field — a phenomenon that generates regular, pulsed electromagnetic signals at frequencies Tesla's equipment could detect. If this explanation is correct, Tesla didn't hear Martians. He inadvertently became the first person to detect radio emissions from another planet's magnetosphere — a discovery that wouldn't be formally made for another fifty-five years.

The honest assessment is that we don't know what Tesla heard. What we know is that his equipment was sensitive enough to detect something that no one else's equipment could detect, that he reported it carefully and with appropriate qualification, and that the press turned it into a carnival. The signals remain an open question — but the receiving apparatus that detected them was, by any measure, the most sensitive electromagnetic instrument on the planet in 1899.

* * *

There is one more result from Colorado Springs that deserves particular attention, because it connects Tesla's experimental work to a discovery that wasn't formally made until more than fifty years later.

Tesla's stationary wave observations led him to calculate that the Earth possessed a resonant frequency in the extremely low range. His estimates, as recorded in his writings from the early 1900s, placed this frequency at approximately six to twelve cycles per second. He understood the mechanism: the Earth's surface and the conducting layer of the upper atmosphere formed a spherical cavity, and electromagnetic waves within that cavity would resonate at frequencies determined by the planet's circumference and the speed of light. As early as 1893, in his lecture "On Light and Other High Frequency Phenomena," Tesla had speculated: "If ever we can ascertain at what period the earth's charge, when disturbed, oscillates with respect to an oppositely electrified system or known circuit, we shall know a fact possibly of the greatest importance to the welfare of the human race."

In Colorado, he went looking for that period. And he found evidence of it.

In 1952, the German physicist Winfried Otto Schumann, working at the Technical University of Munich, mathematically predicted that the Earth-ionosphere cavity should sustain standing electromagnetic waves at a fundamental frequency of approximately 7.83 hertz. The prediction was confirmed by measurements made by Balser and Wagner between 1960 and 1963. These are now called the Schumann resonances, and they represent one of the fundamental electromagnetic properties of the planet.

Tesla's estimate was not precisely correct — his figures of six to twelve hertz bracketed the true value rather than pinpointing it, and there is scholarly debate about whether he was measuring the same phenomenon Schumann later described or a related but distinct form of terrestrial resonance involving conduction currents through the earth's body rather than electromagnetic waves in the atmospheric cavity. The distinction matters technically. But the principle Tesla identified — that the planet possesses natural resonant frequencies, that these can be excited by electrical oscillators, and that a transmitter tuned to match them could achieve extraordinary efficiency in energy transfer — was confirmed by Schumann's work half a century later. Tesla saw the outline of the phenomenon. The details took another fifty years to fill in.

This is worth stating precisely, because it's a place where Tesla enthusiasts tend to overclaim and Tesla skeptics tend to dismiss. Tesla did not discover the Schumann resonance. Schumann did, in 1952. But Tesla, in 1899, experimentally detected evidence that the planet behaves as a resonant electromagnetic system, calculated an approximate frequency for that resonance, and built a transmitter designed to exploit it. His estimate was in the right order of magnitude. His physical reasoning was sound. And the practical consequence he drew from it — that this resonance could be used for wireless power transmission — remains an open engineering question to this day.

* * *

In December 1899, as Tesla's time in Colorado was drawing to a close, a photographer named Dickenson Alley arrived from New York to document the laboratory for *The Century Magazine*. Alley spent two weeks working in freezing conditions, producing sixty-eight photographs using eleven-by-fourteen-inch glass plates with exposures sometimes lasting more than an hour by moonlight.

The most famous image shows Tesla seated calmly in a chair, reading a book, while enormous electrical discharges arc and writhe from the magnifying transmitter behind him. The sparks span twenty-two feet. The photograph is magnificent, and it is a double exposure — Tesla sat for the portrait in a separate exposure from the one that captured the discharge. He was not actually present in the room while the coil was firing. The image was always intended as theatre, not documentation, and Tesla inscribed a print of it to his friend Sir William Crookes with characteristic charm: "To my illustrious friend Sir William Crookes of whom I always think and whose kind letters I never answer!"

The photographs accompanied a twelve-thousand-word essay Tesla wrote for the June 1900 issue of *The Century Magazine*, titled "The Problem of Increasing Human Energy." It remains one of the most remarkable articles ever published in a popular magazine. Tesla ranged freely across wireless power transmission, renewable energy, the harnessing of solar and wind power, global demilitarization through a form of mutually assured destruction, the liberation of humanity from dependence on fossil fuels, and a theory of consciousness that bears resemblance to what would later be called the many-worlds interpretation. At the center of his vision was the magnifying transmitter — perfected, scaled up, and deployed as the backbone of a world wireless system that would deliver energy and information to every point on the planet.

It was, depending on your perspective, either the most visionary engineering proposal of the nineteenth century or the most grandiose fantasy. Tesla believed it was engineering. History has not yet delivered a final verdict.

* * *

Tesla left Colorado Springs on January 7, 1900. He'd been there for just under eight months. He left behind unpaid bills for electricity and water, a caretaker's wages in arrears, and a laboratory full of equipment. The locals expected him to return. He never did.

When Tesla refused to settle his debts, the laboratory and its contents were eventually seized and sold to a man named C.E. Maddocks, who used the lumber to build a house. By 1904, the most advanced electromagnetic research facility in the world had been reduced to timber and nails. Nothing remains at the site today except a small commemorative plaque.

Tesla returned to New York broke but exhilarated. He was convinced that he'd performed the most important experiments of his life. He'd proven that the Earth was a resonant conductor. He'd transmitted power through the ground without wires. He'd produced the largest artificial lightning discharges in history and shown that resonant amplification could generate millions of volts from a relatively modest input. He'd mapped the electrical character of the planet with instruments of unprecedented sensitivity. And he'd conceived, in full technical detail, a system that could — he believed — deliver wireless power and wireless communication to every point on the globe.

All he needed now was money. And there was exactly one man alive with enough of it, and enough appetite for transformative technology, to fund what Tesla had in mind.

His name was J.P. Morgan.

* * *

Colorado Springs was the apex of the resonance idea. Everything Tesla had done — from the rotating magnetic field to the polyphase system to the Tesla coil to the high-frequency lectures — had been building toward this. The principle was always the same: identify the natural frequency of a system and drive it in rhythm. At the scale of a motor, this produced rotation. At the scale of a transformer, it produced amplification. At the scale of a lecture hall, it produced visible, theatrical electricity. At the scale of a power grid, it lit cities.

At the scale of a planet, Tesla believed, it could power civilization.

Whether he was right about that — whether the magnifying transmitter could actually have delivered on the promise of global wireless power — is a question that has never been definitively answered, because the full-scale system was never built. What was built, at Colorado Springs, was a proof of concept that demonstrated every component of the idea in isolation: resonant amplification, wireless energy transfer through the earth, detection of the planet's electromagnetic characteristics, and the generation of electrical effects at scales previously achieved only by nature. Each of these was real. Each was documented in Tesla's diary with meticulous precision. The question that remains is whether they could have been integrated into a working global system, or whether the losses and complexities of planetary-scale transmission would have defeated even Tesla's extraordinary engineering.

Tesla didn't think there was any question at all. He'd felt the planet respond. He'd tuned his transmitter to the earth and the earth had rung. The next step was obvious, inevitable, and — he was certain — achievable.

He just had to build a bigger one.

CHAPTER NINE

THE TOWER

Wardenclyffe, 1900–1917

My Inventions

By Nikola Tesla

V. The Magnifying Transmitter

AS I review the events of my past life I realize how subtle are the influences that shape our destinies. An incident of my youth may serve to illustrate. One winter's day I managed to climb a steep mountain, in company with other boys. The snow was quite deep and a warm southerly wind made it just suitable for our purpose. We amused ourselves by throwing balls which would roll down a certain distance, gathering more or less snow, and we tried to outdo one another in this exciting sport. Suddenly a ball was seen to go beyond the limit, swelling to enormous proportions until it became as big as a house and plunged thundering into the valley below with a force that made the ground tremble. I looked on spellbound, incapable of understanding what had happened. For weeks afterward the picture of the avalanche was before my eyes and I wondered how anything so small could grow to such an immense size. Ever since that time the magnification of feeble actions fascinated me, and when, years later, I took up the experimental study of mechanical and electrical resonance, I was keenly interested from the very start. Possibly, had it not been for that early powerful impression, I might not have followed up the little spark I obtained with my coil and never developed my best invention, the true history of which I will tell here for the first time.

This Photograph Shows the Famous Tesla Tower Erected at Shoreham, L. I., N. Y. The Tower Was Dismantled at the Outbreak of the War. It Was 187 Feet High. The Spherical Top Was 68 Feet in Diameter.

Note the Huge Size of the Structure by Comparing the Two-story Power Plant in the Rear. The Tower Which Was to be Used by Tesla in His "World Wireless," Was Never Finished. Illustration Opposite Shows It Completed.

Scrapping the World's Engines.

"Lionhunters" have often asked me which of my discoveries I prize most. This depends on the point of view. Not a few technical men, very able in their special departments, but dominated by a pedantic spirit and near-sighted, have asserted that excepting the induction motor I have given to the world little of practical use. This is a grievous mistake. A new idea must not be judged by its immediate results. My alternating system of power transmission came at a psychological moment, as a long-sought answer to pressing industrial questions, and altho considerable resistance had to be overcome and opposing interests reconciled, as usual, the commercial introduction could not be long delayed. Now, compare this situation with that confronting my turbine, for example. One should think that so simple and beautiful an invention, possessing many features of an ideal motor, should be adopted at once and, undoubtedly, it would under similar conditions. But the prospective effect of the rotating field was not to render worthless existing machinery; on the contrary, it was to give it additional value. The system lent itself to new enterprise as well as to improvement of the old. My turbine is an advance of a character entirely different. It is a radical departure in the sense that its success would mean the abandonment of the antiquated types of prime movers on which billions of dollars have been spent. Under such circumstances the progress must needs be slow and perhaps the greatest impediment is encountered in the prejudicial opinions created in the minds of experts by organized opposition. Only the other day I had a disheartening experience when I met my friend and former assistant, Charles F. Scott, now professor of Electrical Engineering at Yale. I had not seen him for a long time and was glad to have an opportunity for a little chat at my office. Our conversation naturally enough drifted on my turbine and I became heated to a high degree. "Scott," I exclaimed, carried away by the vision of a glorious future, "my turbine will scrap all the heat-engines in the world." Scott stroked his chin and looked away thoughtfully, as though making a mental calculation. "That will make quite a pile of scrap," he said, and left without another word!

IMAGINE a man a century ago, bold enough to design and actually build a huge tower with which to transmit the human voice, music, pictures, press news and even power, thru the earth to any distance whatever without wires! He probably would have been hung or burnt at the stake. So when Tesla built his famous tower on Long Island he was a hundred years ahead of his time. And foolish ridicule by our latter day arm-chair "savants," does not in the least mar Tesla's greatness.

The titanic brain of Tesla has hardly produced a more amazing wonder than this "magnifying transmitter." Contrary to popular belief his tower was not built to radiate Hertzian waves into the ether. Tesla's system sends out thousands of horsepower thru the earth—he has shown experimentally how power can be sent without wires over distances from a central point. Nor is there any mystery about it how he accomplishes the result. His historic U. S. patents and articles describe the method used. Tesla's Magnifying Transmitter is truly a modern lamp of Aladdin.

EDITOR.

"Aladdin's Lamp".

These and other inventions of mine, however, were nothing more than steps forward in certain directions. In evolving them I simply followed the inborn instinct to improve the present devices without

(Continued on page 148)

Tesla returned to New York in January 1900 with no money and absolute certainty. He'd left Colorado Springs in a condition that would become characteristic of the second half of his life: broke, exhilarated, and carrying plans so vast that no reasonable person could fund them and no unreasonable person could resist hearing about them. He owed money for electricity, water, and the wages of a caretaker he'd promised to pay when he came back. He never came back.

What he carried instead was data. Five hundred pages of meticulously recorded notes, two hundred drawings, and the experimental proof — demonstrated, measured, and documented — that the Earth behaves as a resonant conductor. He'd transmitted power through the ground without wires. He'd produced the largest artificial lightning in history. He'd measured the electrical character of the planet with a sensitivity no one had previously attempted. And he'd conceived, in full technical detail, a system for doing all of this at global scale: wireless power, wireless communication, wireless everything, delivered through the body of the planet to any point on its surface.

He called it the World System. Now he needed someone to pay for it.

* * *

Before leaving Colorado, Tesla had agreed to write an article for *The Century Magazine*, one of the most prestigious publications in America. Robert Underwood Johnson, the editor and a close friend, had sent the photographer Dickenson Alley to Colorado Springs to document the experiments, and now he wanted the story. Johnson specifically asked Tesla not to write a metaphysical treatise. Tesla, being Tesla, wrote a twelve-thousand-word metaphysical treatise.

"The Problem of Increasing Human Energy," published in the June 1900 issue, is one of the strangest documents in the history of science. It contains, threaded through passages of startling prescience, Tesla's vision of wireless power transmission, renewable energy from solar and wind sources, global demilitarization through a doctrine resembling mutually assured destruction, liberation from fossil fuels, and a theory of consciousness that anticipates certain features of the many-worlds interpretation. It describes the magnifying transmitter as the backbone of a global wireless system. It champions a future in which machines do the dangerous work and humanity is freed to pursue higher aims. And it is accompanied by Alley's extraordinary photographs of the Colorado Springs apparatus in action, including the famous double exposure of Tesla reading calmly while twenty-two-foot lightning bolts arc above his head.

The article galvanized public opinion. People read it as visionary. The scientific community read it as unhinged. But it caught the attention of one particular reader: John Pierpont Morgan, the most powerful financier in America, a man whose personal fortune and web of corporate interests gave him effective control over much of the nation's banking, steel, and railroad industries. Morgan had already invested in Niagara. He understood, at least commercially, what Tesla's polyphase system had accomplished. And one detail in the *Century* article interested him specifically: a magnifying transmitter capable of locating ships at sea and relaying their position to the mainland. Morgan was a yachtsman. His own negotiations with Marconi for a secure wireless communication system had failed. Tesla's alternative intrigued him.

* * *

Tesla spent the second half of 1900 courting investors. He wined and dined potential backers at the Waldorf-Astoria's Palm Garden, where he was now living, and at the Players Club and Delmonico's. He approached his old ally George Westinghouse first. Westinghouse declined to invest but lent Tesla six thousand dollars and suggested he try the venture capitalists. Tesla talked to John Jacob Astor, Thomas Fortune Ryan, and others. Most listened politely and wrote nothing.

Morgan was different. The discussions between Tesla and Morgan, facilitated in part through the social circles of Robert and Katharine Johnson, extended over several months. Tesla described a wireless telegraphy system that would transmit messages across the Atlantic, relay telephonic signals, and even send facsimile images — all superior to Marconi's crude single-frequency apparatus. Morgan, who had a banker's instinct for monopoly, saw in Tesla's patents the possibility of controlling the next generation of transatlantic communication. In March 1901, he wrote Tesla a cheque for a hundred and fifty thousand dollars.

The terms were precise. In return for the investment, Morgan received a fifty-one per cent interest in Tesla's patents and any future inventions relating to wireless telegraphy, wireless telephony, and electric lighting. Not a partnership. Not a venture fund. A controlling share. Morgan would own the majority of everything Tesla invented in the field. For Tesla, the hundred and fifty thousand dollars was a beginning. For Morgan, it was the whole thing.

This mismatch would destroy the project.

* * *

The contract signed on March 1, 1901, specified a wireless telegraphy system. Morgan understood he was funding a radio station — a more sophisticated competitor to Marconi's transatlantic telegraph, capable of secure multichannel communication. This was commercially rational. Marconi was already demonstrating wireless telegraphy over increasing distances, and Morgan, who had extensive shipping and financial interests on both sides of the Atlantic, could see the value in owning the superior system.

Tesla was building something else.

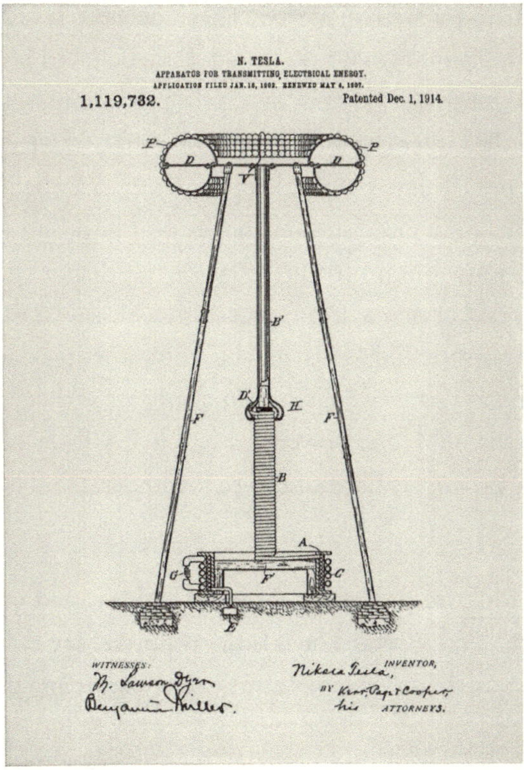

What Tesla had designed, and what the Colorado Springs experiments had been conducted to validate, was not a radio station. It was a planetary resonance machine. The system Tesla intended to construct was the full-scale implementation of the magnifying transmitter principle: a facility that would drive electrical currents into the earth at its resonant frequency, establish standing waves across the entire planet, and deliver power — not just signals, but usable electrical power — to any point on the surface through tuned receivers. The wireless telegraphy component was real and would work. But it was the smallest feature of a system designed to replace the entire electrical grid.

Tesla knew this when he signed the contract. The question of whether he deliberately concealed the full scope from Morgan or simply failed to communicate it adequately has been debated by every biographer since. The charitable interpretation is that Tesla believed he could demonstrate the telegraphy system first, prove its commercial viability, and then reveal the larger capability once Morgan had seen returns on his investment. The less charitable interpretation is that Tesla took money earmarked for a radio station and spent it building a power plant. Both interpretations contain truth. Tesla was not dishonest, but he was catastrophically bad at managing expectations. He saw the telegraphy and the power transmission as inseparable components of a single system — which, technically, they were. The same apparatus that sent signals through the earth could send power through it. To Tesla, building the one without the other was like building a locomotive and declining to lay track.

Morgan didn't see it that way. Morgan had invested in a telegraph. When he eventually discovered he'd funded a project to deliver free electricity to the world, his response was exactly what any banker's would have been.

* * *

Tesla needed land: a large, flat tract with good ground conductivity, access to a railway for transporting heavy equipment, and enough distance from Manhattan that his experiments wouldn't alarm the population. He found it through James S. Warden, a western lawyer and banker who had purchased land on Long Island's north shore and was developing a resort community he called Wardenclyffe-on-Sound. Warden saw Tesla's project as the nucleus of a future industrial city — a "Radio City" that would employ thousands of workers who'd build homes on his adjacent land. He offered Tesla two hundred acres, sixty-five miles from New York, near the village of Shoreham, with railway access and proximity to Long Island Sound.

Tesla accepted. He hired his friend Stanford White — of the architectural firm McKim, Mead & White, and one of the most celebrated architects in America — to design the laboratory building. White produced an elegant red-brick structure, ninety-four feet square, in the Italian Renaissance style. It housed a laboratory, machine shop, boiler room, generator room, instrumentation room, a library, and an office. Inside were Tesla coils, X-ray equipment, glass-blowing apparatus, electrical generators and transformers, and the remote-controlled boat he'd demonstrated at Madison Square Garden in 1898. It was, by any measure, one of the finest private research facilities in the world.

The tower was designed by W.D. Crow, an associate of White's. It was an extraordinary structure: a wooden-framed octagonal column rising a hundred and eighty-seven feet into the air, topped by a hemispherical cupola sixty-eight feet in diameter. The cupola was a steel framework weighing fifty-five tons, designed to be clad in copper and function as an enormous elevated capacitor — the same role the thirty-inch copper sphere had played at Colorado Springs, but on a vastly larger scale. Every structural element was designed to be individually removable and replaceable, giving the tower an almost modular character unusual for the period.

But the most expensive and most important part of the facility was invisible. Directly beneath the tower, Tesla sank a shaft — ten by twelve feet, lined with timber and steel — a hundred and twenty feet straight down into the Long Island earth. A spiral staircase descended to the bottom. From the base of the shaft, sixteen iron pipes were driven outward through the subsoil, each pushed three hundred feet by special machines Tesla designed for the purpose. Four tunnels, possibly brick-lined and waterproofed, radiated from the shaft's base in the cardinal directions, extending a hundred feet each.

Tesla explained the function of this underground network with characteristic directness. "In this system that I have invented," he said, "it is necessary for the machine to get a grip of the earth, otherwise it cannot shake the earth. It has to have a grip on the earth so that the whole of this globe can quiver."

The shaft, the pipes, the tunnels — these were the ground connection. At Colorado Springs, Tesla had used a copper plate buried twelve feet deep, and even that modest connection had been enough to transmit power through the earth to receivers twenty-six miles away. The Wardenclyffe ground system was engineered at a completely different scale: iron pipes driven three hundred feet into glacial sediment, reaching toward the water table, designed to couple the transmitter's output into the body of the planet with minimal resistance. If the copper sphere at the top of the tower was one plate of a planetary-scale capacitor, the underground network was the other. Everything between them — tower, coils, resonant circuits — existed to pump energy between these two terminals at exactly the right frequency.

This was not a radio antenna. Marconi's system radiated electromagnetic waves into the air and hoped a receiver would catch a tiny fraction of them. Tesla's system drove current into the earth and expected the planet to deliver it. The distinction is everything. And it's the distinction Morgan never understood — or, if he understood it, never accepted.

* * *

What, exactly, did Tesla intend Wardenclyffe to do? He laid out the full vision in a series of articles and prospectuses between 1900 and 1904, most notably in "The Transmission of Electrical Energy Without Wires," published in *Electrical World and Engineer* on March 5, 1904. The World System, as he described it, would consist of a network of transmitting stations, each located near a major center of civilization, connected not by wires but by the earth itself. The services this network would provide included:

The interconnection of all existing telephone exchanges and telegraph stations worldwide. The transmission of news, music, stock market reports, private messages, and government communications to any point on the planet. The universal dissemination of information through a receiver no larger than a watch, cheap enough for anyone to carry. Facsimile transmission of documents and images. A universal system of marine navigation, including synchronization of precision timepieces. Secure military communications. Remote control and guidance of unmanned vehicles. And the wireless transmission of electrical power for industrial purposes, for lighting, and for the propulsion of ships and aircraft.

Read that list carefully. Navigation by signal synchronization is GPS. A pocket receiver for news and messages is a smartphone. Facsimile transmission is the internet. Universal interconnection of telephone exchanges is the global telecommunications network. Tesla described, in a technical prospectus published in 1904, substantially every major feature of the communications infrastructure that exists today — and one feature, wireless power transmission, that still doesn't.

The critical difference between Tesla's World System and what eventually emerged is the medium. Tesla's network would operate through the earth. The system that actually developed over the following century — radio, television, satellite communication, cellular networks, the internet — operates through the air, using exactly the Hertzian radiation Tesla believed was wasteful and inferior. Whether Tesla's alternative would have worked at global scale is a question that has never been answered, because the only facility designed to test it was never completed.

* * *

Construction began in September 1901. From the start, money was a problem.

Morgan had committed a hundred and fifty thousand dollars, but the full amount was not delivered at once. Tesla complained in letters to Stanford White that he was already facing foreclosure before the foundations were complete. The underground work alone — the shaft, the tunnels, the iron pipes driven hundreds of feet through resistant soil by specially designed machines — consumed an enormous portion of the budget. Tesla had designed the most expensive part of the facility to be the part no one could see.

In March 1901, he'd contracted with the Westinghouse Electric Company to manufacture the generators and transformers for the transmitter. Two massive units were assembled: one for Wardenclyffe, and a second warehoused for future delivery to a planned receiving station across the Atlantic. The Westinghouse engineers worked closely with Tesla on the design, but the equipment was expensive, and every component had to be custom-built. The two-hundred-kilowatt alternating-current industrial generator that would power the system was, by itself, a significant capital investment.

Then, in July 1901, Tesla made the decision that would doom the project. He instructed his engineers to scale up the facility beyond the original specifications. The tower was to be taller, the transmitter more powerful, the ground system deeper. The revised design was not the wireless telegraphy station Morgan had funded. It was the full World System transmitter — a planetary resonance machine designed to demonstrate not just communication but power transmission. Tesla made this decision because Marconi was closing in.

* * *

Guglielmo Marconi had been transmitting wireless signals over increasing distances since the late 1890s. His system was crude compared to Tesla's — a single-frequency spark transmitter using Hertzian radiation, without the resonant tuning, multichannel capability, or ground-conduction mechanism that Tesla's patents described. It was also, in significant part, built on Tesla's own work. Tesla would later claim that seventeen of his patents made Marconi's transatlantic transmission possible, and the claim was not empty: Marconi had been denied a U.S. patent in 1900 precisely because Tesla's prior art anticipated the key innovations. The Patent Office reversed itself in 1904 and granted Marconi patent number 763,772, but in 1943 — nine months after Tesla's death — the United States Supreme Court would invalidate Marconi's broad claims, ruling that they had been anticipated by Tesla's earlier patents. The vindication came too late to matter, and it received almost no publicity. But in 1901, none of that had happened yet. What had happened was that Marconi possessed the one advantage that outweighed every technical deficiency: he was building, testing, and demonstrating, while Tesla was designing, planning, and fundraising.

In December 1901, Marconi announced that he had transmitted the Morse code letter S — three dots — across the Atlantic Ocean, from Poldhu in Cornwall to Signal Hill in St. John's, Newfoundland. The claim was disputed by some, and the exact nature of what Marconi's receiver detected remains debated. But the announcement was a sensation. The press declared the age of wireless communication had arrived, and the man who'd ushered it in was Marconi, not Tesla.

Tesla's response was to build bigger. If Marconi could send three dots across the Atlantic with a crude single-frequency transmitter, Tesla would build a system that could send everything — voice, data, images, power — through the earth itself. The scale of Wardenclyffe had to increase to make the demonstration so overwhelming that no one would mistake who held the superior technology. This was sound engineering logic and catastrophic financial judgment.

Tesla wrote to Morgan, explaining that the facility needed to be expanded, that the original budget was insufficient, and that the redesigned system would be so superior to Marconi's as to render it commercially obsolete. Morgan's reply was silence, and then refusal. The contract specified a wireless telegraphy station. Morgan had funded a specific project with a specific budget. He had not agreed to fund an open-ended research programme into planetary-scale power transmission.

* * *

What followed was one of the most painful correspondences in the history of American technology. Between 1901 and 1904, Tesla wrote Morgan letter after letter — pleading, explaining, reassuring, and increasingly desperate. He assured Morgan that the system would be superior to Marconi's. He argued that the investment would be recovered many times over. He described the specific commercial applications: transatlantic telegraphy, maritime communication, news broadcasting. He downplayed the power transmission component when he sensed Morgan's resistance, then brought it forward again when he thought a grander vision might rekindle interest.

Morgan was unmoved. He had made his investment, defined its scope, and had no intention of expanding it. He wrote to Tesla with the blunt clarity of a man who controlled more capital than most nations: "I have received your letter, and in reply would say that I should not feel disposed at present to make any further advances."

Tesla also faced an additional financial blow he hadn't anticipated. The stock market panic of 1901, triggered in part by a battle for control of the Northern Pacific Railway in which Morgan himself was a principal combatant, drove up the cost of raw materials across the board. Steel, copper, lumber — everything Tesla needed for construction became significantly more expensive. The hundred and fifty thousand dollars, already inadequate for the expanded design, now bought substantially less than it had when the contract was signed.

Construction continued through 1902 despite the financial crisis. In June, Tesla began moving his laboratory operations from his Houston Street workshop in Manhattan to the partially completed Wardenclyffe facility. By the end of 1902, the tower had reached its full height of a hundred and eighty-seven feet. The cupola framework was in place atop it. The laboratory building was functional. The underground shaft had been sunk and the iron pipes driven into the earth. The facility was, structurally, very nearly complete.

What it lacked was the transmitting apparatus. The tower was a skeleton. The cupola's steel framework had never been clad in its intended copper sheathing. The resonant coils, the driving oscillators, the control systems — the equipment that would have made the tower a functioning magnifying transmitter rather than an impressive piece of architecture — were incomplete. The building was finished. The machine inside it was not.

* * *

In July 1903, Tesla ran a partial test of the system. The details are sparse — Tesla kept the public and press away from Wardenclyffe, and what is known comes from fragmentary reports and later accounts. But what the surrounding population witnessed was spectacular. Brilliant flashes of electrical discharge erupted from the tower, visible for miles in every direction, reportedly seen as far away as Connecticut. The display lasted long enough to generate newspaper coverage and considerable local alarm. No explanation was offered by Tesla or his staff.

The test, whatever it proved to Tesla, proved nothing to Morgan. The tower never operated at full power again during the period of active development. Whether this was because the partial test revealed technical problems, because funds had been exhausted, or because Tesla deliberately limited further testing until the apparatus could be completed is unknown. What is known is that after 1903, the pace of work at Wardenclyffe slowed and then stopped.

* * *

By July 1904, Morgan and the other investors had made their final decision: no additional financing would be provided. Worse, Morgan actively discouraged other potential investors from supporting the project. When Tesla approached new backers, he found doors closed that should have been open. Morgan's influence in financial circles was absolute, and his withdrawal from Wardenclyffe carried an implicit message: the project was not worth funding.

Tesla tried everything. He issued a prospectus describing the World System in detail, hoping to attract independent investment. He published the "Tesla Manifesto" in *Electrical World and Engineer* in February 1904, a four-page circular printed on vellum, sealed with red wax stamped "N.T.," listing ninety-three U.S. patents and announcing his availability as a consulting engineer. He wrote to his secretary George Scherff with the exhausted candor of a man who had run out of maneuverers: "I swear! If I ever get out of this hole nobody will catch me without cash."

In May 1905, Tesla's patents on alternating current motors and other methods of power transmission expired. The royalty payments that had provided his primary income for years stopped overnight. The financial blow was devastating. The man whose inventions powered cities across the world no longer earned a cent from them.

Workers were laid off. Stanford White, Tesla's friend and the architect of the Wardenclyffe building, was murdered in June 1906 by Harry Kendall Thaw, the husband of White's former lover Evelyn Nesbit, in what became one of the most sensational crimes of the era. The building stood. The tower stood. But the life had gone out of both.

Tesla suffered what contemporaries described as a nervous breakdown. The man who had lit cities, who had held audiences spellbound with fountains of electrical fire, who had made Niagara fall obediently into his polyphase system, was now unable to pay his bills, unable to fund his research, unable to attract investment, and unable to complete the single project that he believed would vindicate everything he'd ever done.

* * *

By 1906, Wardenclyffe was effectively abandoned. Parts of the building remained in use until 1907, but the tower stood empty — a skeleton against the Long Island sky, its cupola unclad, its transmitter incomplete, its underground network of pipes and tunnels silent. The most advanced electromagnetic research facility in the world had become a monument to what might have been.

In March 1916, the publication *Export American Industries* ran a story titled "Tesla's Million Dollar Folly" describing the abandoned site. The reporter found everything left as if for a single day's absence: chairs, desks, and papers in businesslike array, the great wheels seeming to await Monday morning. But the magic word had not been spoken, and no one was coming to speak it.

Tesla had taken out a mortgage on the Wardenclyffe property in 1904, using it as collateral to cover his living expenses at the Waldorf-Astoria. The mortgage was held by George C. Boldt, the hotel's proprietor. In 1908, Tesla took out a second mortgage from Boldt. By 1915, his accumulated debt at the hotel was approximately twenty thousand dollars. When Tesla could make no further payments, Boldt foreclosed.

There is a further cruelty in this detail that deserves noting. Tesla's living arrangements at the Waldorf-Astoria had originally been facilitated by his friend and early backer John Jacob Astor IV. Astor had extended

Tesla credit at the hotel and supported his work with personal investment. But Astor perished in the sinking of the *Titanic* in April 1912, and his heirs required payment of the debts. The man who had believed in Tesla enough to fund him was dead, and the obligations he'd underwritten now became the instrument of Tesla's dispossession.

* * *

On July 4, 1917, the Smiley Steel Company of New York arrived at Wardenclyffe with dynamite. The United States had entered the First World War three months earlier, and newspapers reported that the tower was being demolished on government orders — that it was suspected of being used as a navigational landmark for German submarines, or as a communications station for enemy agents. The rumor persisted for decades.

The truth was simpler and worse. The tower was demolished to pay Tesla's hotel bill. Boldt had foreclosed on the property and ordered the structure taken down for scrap. The initial dynamite blast knocked the tower to a tilt but didn't bring it down completely. Further demolition continued through September before the structure was fully dismantled. The scrap value of the steel realized seventeen hundred and fifty dollars.

Seventeen hundred and fifty dollars. The most ambitious electromagnetic engineering project ever conceived, reduced to less than two thousand dollars in scrap metal. The facility that was to have been the first node in a global wireless network — the prototype for a system Tesla envisioned spanning thirty stations worldwide, delivering power and information to every point on the planet — was dynamited on Independence Day and sold for parts.

Tesla was not in New York for the demolition. He could not have stopped it. On April 20, 1922, he lost an appeal of judgment in Boldt's foreclosure, which locked him out of any future claim on the property. The land was transferred to new owners. Within a few years, the site was leased to Peerless Photo Products, which later became part of AGFA Corporation. For the next fifty years, the laboratory designed by Stanford White and the underground tunnels designed by Nikola Tesla were used for manufacturing photographic supplies.

* * *

The standard narrative of Wardenclyffe, told in virtually every popular account, is a story of hubris: Tesla overreached, his ambitions exceeded his resources, and the project collapsed under the weight of its own grandiosity. There is enough truth in this to make it satisfying and enough omission to make it misleading.

Tesla's financial management was genuinely terrible. He took a fixed sum of money earmarked for a specific project and spent it on a different, vastly more expensive project without securing additional funding first. He made engineering decisions driven by the desire to outcompete Marconi rather than by the budget he actually had. He failed to communicate the true scope of his plans to his primary investor, then was shocked when that investor declined to fund the scope he'd concealed. These are real failures. Any honest account has to acknowledge them.

But the hubris narrative obscures what was actually at stake. The question was not whether Tesla was a good businessman. He was a dreadful businessman, and everyone who knew him knew it. The question was whether the physics worked. Whether a properly constructed magnifying transmitter, grounded deep in the earth and driving current at the planet's resonant frequency, could establish standing waves capable of delivering usable power at distances of thousands of miles. Tesla had demonstrated every component of this system in isolation at Colorado Springs. He had transmitted power through the ground. He had measured the earth's resonant response. He had calculated the Q factor that determined whether the system was feasible or futile. What he had not done — what Wardenclyffe was designed to do — was integrate all of these components into a single working system operating at the scale necessary to prove global viability.

That test never happened. Not because the physics failed, but because the money stopped.

* * *

This demands careful separation. There are three distinct questions tangled together in the Wardenclyffe story, and they have three different answers.

First: could Tesla's system transmit wireless signals through the earth? Yes. He did it at Colorado Springs. He did it at twenty-six miles. The physics of ground-wave propagation is well established, and Tesla's Colorado demonstrations are not seriously disputed by modern electrical engineers.

Second: could a Wardenclyffe-scale transmitter have established resonant standing waves in the earth? Probably. The earth does behave as a resonant electromagnetic cavity — this was confirmed by Schumann in 1952 and measured by Balser and Wagner in the early 1960s. Tesla's estimate of the planet's resonant frequency was approximately correct. Whether the specific Wardenclyffe apparatus could have excited this resonance at useful amplitude depends on engineering details that were never tested, but the underlying principle is not in dispute.

Third: could the system have delivered usable electrical power at global distances through the earth with acceptable efficiency? This is where honest assessment requires honesty. Nobody knows. The feasibility depends on the ratio of energy input to energy recovered at the receiver, which depends on the Q factor of the earth-ionosphere cavity at the operating frequency, the coupling efficiency of the ground connection, the losses in the transmission medium, and the practical engineering of the receivers. Tesla believed his Colorado measurements showed the Q factor was high enough. Modern analysis of Tesla's data by the Corum brothers and others suggests the numbers were in the plausible range. But plausible is not proven. The full-scale test was never conducted.

The point is not that the question was never answered. The point is that Tesla had spent his entire career answering questions like this, and he had never once been wrong. The rotating magnetic field worked. The polyphase system worked. The Tesla coil worked. The magnifying transmitter worked. Colorado Springs worked. Every theory Tesla developed resulted in a working machine, every prediction he made about resonance was confirmed by experiment, and every apparatus he designed performed as calculated. He was not Edison, trying ten thousand things to find one that worked. Tesla worked things out in his mind, built them once, and they worked. That was his method and his record, without exception, across four decades of engineering.

Wardenclyffe was not an experiment in the conventional sense. Tesla did not build a tower of that complexity — a shaft sunk a hundred and twenty feet into the earth, iron pipes driven three hundred feet through subsoil, four tunnels radiating in cardinal directions, a fifty-five-ton steel cupola — to find out whether the earth would resonate. He had already proven that at Colorado Springs. He built Wardenclyffe to demonstrate, at full planetary scale, something he already knew was true. It was not a question to be answered. It was a machine to be operated. And it was demolished for scrap before it could be switched on.

* * *

Morgan has been cast as the villain of the Wardenclyffe story, and there is a popular version in which his refusal to fund Tesla was motivated by the realization that wireless power couldn't be metered — that you couldn't charge for electricity that flowed freely through the earth. "Where can I put the meter?" Morgan is supposed to have asked. The line has the ring of apocrypha, and its attribution is uncertain, but it captures a real tension.

Morgan's logic, however, doesn't require malice to explain. He had invested a hundred and fifty thousand dollars in a wireless telegraphy station. Tesla had used the money to begin constructing something far more ambitious and far more expensive, without disclosing the change in scope or securing the additional funding it required. From Morgan's perspective, the inventor had violated the terms of their agreement. When Tesla asked for more money, Morgan had every commercial reason to refuse. The project was over budget, behind schedule, and had been redesigned without the investor's consent. No venture capitalist in history has responded well to that combination.

At the same time, Morgan's decision to actively discourage other investors went beyond protecting his own position. When Morgan told others not to fund Wardenclyffe, he was not merely declining further involvement. He was killing the project. In a financial ecosystem where Morgan's word carried the force of law, his withdrawal was a death sentence. Tesla could not have found alternative funding from any major

American financier once Morgan had signaled his disapproval. The effect was the same as if Morgan had ordered the tower demolished himself. It just took sixteen years longer.

* * *

Every chapter of this book has traced the same principle through increasing scale. The rotating magnetic field in a motor. The polyphase system in a power grid. The Tesla coil on a laboratory bench. The magnifying transmitter in a barn on the Colorado prairie. At each stage, the idea was the same: find the natural frequency, drive in rhythm, and the system amplifies the input beyond all proportion to the energy supplied. Resonance. The most powerful principle Tesla ever discovered, applied to progressively larger systems.

Wardenclyffe was the final stage. The system in which the resonant body was the planet itself. Everything Tesla had built pointed toward this moment: the demonstration that the earth could be set vibrating like a bell, that its energy could be tapped at any point on its surface, that the entire infrastructure of wires and cables and generating stations that electrified the modern world could be replaced by a handful of transmitting towers driving current into the ground at the right frequency. It was the culmination of a career in which every preceding stage had worked exactly as designed. Nobody who understood Tesla's track record had reason to doubt that this one would work too. What they doubted — with some justification — was whether anyone would pay for it.

Tesla never abandoned the idea. For the remaining four decades of his life, through poverty, isolation, and the slow erosion of his reputation, he continued to insist that the World System was technically sound, that only the funding had failed, that the physics was proven and the engineering was achievable. He wrote about it, lectured about it, and proposed it to anyone who would listen. He told attorney Drury W. Cooper in 1916 that he had successfully transmitted and received signals at a distance of about ten miles using the earth as conductor. He never stopped believing that the experiment at Wardenclyffe, if completed, would have changed the world.

Whether the completed machine would have delivered on that promise is a question the twenty-first century has not yet bothered to investigate. Given the record of the man who built it, that silence is not caution. It is negligence.

* * *

The Stanford White building still stands. It is the only surviving laboratory of Nikola Tesla. For decades after the tower's demolition, it housed a photography supply company, accumulated hazardous waste, and deteriorated. In 2012, a grassroots campaign organized by Jane Alcorn of the Tesla Science Center at Wardenclyffe and internet cartoonist Matthew Inman raised nearly 1.4 million dollars through crowdfunding to purchase the sixteen-acre site. Elon Musk, whose electric car company bears Tesla's name, contributed to the effort. The property was listed on the National Register of Historic Places in 2018. Plans for a Tesla museum are underway.

In 2017, a century after the tower's demolition, a film crew used ground-penetrating radar at the Wardenclyffe site. They confirmed the existence of a series of tunnels stretching for hundreds of feet beneath the property — the underground network Tesla had built to give his machine a grip on the earth. The tunnels are still there. Their original purpose, according to the Tesla Science Center, "remains a mystery to this day."

It is not a mystery. Tesla explained it with perfect clarity, under oath, in foreclosure proceedings in 1923. The shaft went down a hundred and twenty feet. The pipes went three hundred feet further. The current through those pipes was meant to take hold of the earth. The whole of this globe was meant to quiver.

It never got the chance.

CHAPTER TEN

THE LONG FREQUENCY

New York, 1906–1943

The last thirty-seven years of Nikola Tesla's life are the hardest part of this story to tell honestly. The popular version compresses them into a parable of genius undone by madness: the great inventor loses his money, loses his mind, and dies alone in a hotel room feeding pigeons. It's tidy, poignant, and almost entirely wrong. Tesla didn't lose his mind. He lost his funding. What he did with the decades that followed — some of it brilliant, some of it questionable, much of it impossible to evaluate because the records are incomplete — deserves better than a story about pigeons.

* * *

After Wardenclyffe's effective abandonment in 1906, Tesla did what he'd always done: he went back to the laboratory and invented something new. The bladeless turbine, which he'd been developing since at least 1906 and patented in 1913, was a genuinely original piece of engineering. Instead of using blades to extract energy from a moving fluid — the principle behind every conventional turbine — Tesla's design used a stack of smooth, flat discs mounted on a shaft. The fluid entered at the edge and spiraled inward through the narrow gaps between the discs, transferring its energy through viscous drag and molecular adhesion. No blades meant no blade erosion, no blade failure, and no damage to delicate materials passing through the pump.

The principle was sound and the design was elegant. Julius Czito, the son of Tesla's long-time machinist Kolman Czito, built several prototypes. The first, in 1906, used eight discs six inches in diameter, weighed less than ten pounds, and developed thirty horsepower. A larger model in 1910 reached a hundred horsepower. Tesla described it in 1916 as his best invention to date — a remarkable claim from the man who'd invented the polyphase motor and the magnifying transmitter.

The turbine never achieved commercial adoption in Tesla's lifetime. The disc rotational speeds were so high that the metallurgy of the period couldn't prevent the discs from warping during sustained operation. It was an invention ahead of its materials science, much as the magnifying transmitter was an invention ahead of its funding. The Tesla pump, however, eventually found commercial use beginning in the 1980s, particularly for moving fluids that are abrasive, viscous, or loaded with solids — applications where conventional bladed pumps destroy the very material they're meant to transport. Bioengineers have studied it as a means of pumping blood products with less cellular damage.

Tesla also received patents in this period for a speedometer, a flow meter, a lightning protector, and the valvular conduit — now known as the Tesla valve — a fixed-geometry conduit that allows fluid to flow freely in one direction but resists flow in the other, with no moving parts whatsoever. In 1928, he was granted a patent for a vertical take-off and landing aircraft, a concept that wouldn't become practical for decades. These were not the inventions of a man whose mind had failed. They were the inventions of a man whose money had.

* * *

On the evening of May 18, 1917 — the same year the Wardenclyffe tower was dynamited — the American Institute of Electrical Engineers awarded Tesla its highest honor: the Edison Medal. Tesla had initially refused it. Accepting a medal named after the man who'd built his career on opposing everything Tesla had accomplished was, to put it mildly, unappealing. But Bernard A. Behrend, an engineer who'd long admired Tesla's work, persuaded him to reconsider.

The ceremony was held at the Engineering Societies Building in New York. Behrend delivered the introduction, noting with evident pleasure that it was exactly twenty-nine years to the day and hour since Tesla had first presented his polyphase motor paper to the same institute. The distinguished gentlemen of the committee waited. The auditorium filled with whispers. The guest of honor was not there.

Behrend went looking. He found Tesla in Bryant Park, behind the New York Public Library, standing tall in crisp white tie and tails, covered from head to foot with pigeons. They perched on his head, his shoulders, his outstretched arms. They milled about his feet in a gray mass. Tesla saw Behrend approaching and raised a finger to his lips. The pigeons scattered. Tesla followed Behrend back to the auditorium and delivered a gracious speech.

The image is irresistible, and every Tesla biography uses it. But it is usually offered as evidence of eccentricity edging toward instability. It was nothing of the kind. Tesla had been feeding pigeons for years. He would continue doing so for the rest of his life. It was not a symptom. It was a choice.

* * *

From 1900 onward, Tesla lived in hotels. He never owned a home, never maintained a household, never married. The hotels were his domestic life, and the pattern of his residence in them traces the arc of his financial decline with painful precision.

He lived at the Waldorf-Astoria from 1900 to about 1920, during the years of his greatest public prominence and his most devastating financial collapse. He moved to the St. Regis in 1922, where he brought pigeons into his room, provided basket nests near open windows, and nursed injured birds back to health. The hotel management, contending with the mess, eventually forced him out. He moved to the Hotel Pennsylvania. The same thing happened. He moved to the Hotel Governor Clinton. The same thing happened again. At the Governor Clinton, unable to pay his bill, he offered the management a wooden box that he said contained a working model of his particle beam weapon and warned them never to open it. They complied, hiding it in a storeroom. After Tesla's death it was opened and found to contain old electrical components.

In 1934, the Westinghouse Corporation — the company that had built its empire on Tesla's polyphase patents — reached a settlement with Tesla. Westinghouse would pay him a hundred and twenty-five dollars a month as a consulting fee and cover his hotel rent. The arrangement was characterized publicly as a consultancy. In practice, it was a pension — a recognition, decades late, that the man whose inventions had made the company possible was now destitute. Tesla moved to the Hotel New Yorker, at Eighth Avenue and Thirty-fourth Street, into rooms on the thirty-third floor. He would live there for the remaining nine years of his life.

* * *

The pigeons deserve better than the role they're usually given in this story. They are invariably presented as the definitive evidence that Tesla had lost his grip on reality — the great inventor reduced to communing with birds. This is lazy biography. Tesla was an elderly man, living alone, unmarried, with no family in the country and an ever-shrinking circle of friends and professional contacts. He walked every day to Bryant Park, to the public library, to St. Patrick's Cathedral. He fed the birds. He developed particular attachments to individual pigeons. He nursed sick and injured ones. He spent over two thousand dollars caring for a white pigeon with a broken wing and leg, building a device to support her while she healed.

His biographer John J. O'Neill recorded a conversation in which Tesla spoke about this white pigeon with startling intensity. "I have been feeding pigeons, thousands of them, for years," Tesla said. "Thousands of them, for who can tell —" He described one in particular: pure white, with light gray wing tips. "No matter where I was, that pigeon would find me; when I wanted her I had only to wish and call her and she would come flying to me." And then: "When that pigeon died, something went out of my life. I knew my life's work was finished."

Biographers have speculated endlessly about what this meant. Was it delusion? Displacement? A form of love displaced from human connection onto something safer? Perhaps. Tesla had never had a romantic relationship in his life. He had poured his emotional energy into his work for sixty years, and when the work stopped, something had to receive what remained. That it was a pigeon rather than a person says something about Tesla's isolation, but it says nothing about his sanity. Lonely people form attachments to animals. It's not a diagnosis.

* * *

Beginning in the late 1920s, Tesla developed a custom of holding press conferences around his birthday each July. Reporters would visit him at his hotel, and he would make announcements about his current work and future plans. These conferences are the source of both Tesla's most visionary public statements and his most troubling ones.

In 1931, on the occasion of his seventy-fifth birthday, he appeared on the cover of *Time* magazine. The accompanying article treated him as an elder statesman of invention. But as the years passed, the announcements grew increasingly grandiose, and the gap between what Tesla claimed and what he could demonstrate widened. He described cosmic ray motors that could harness energy from space. He spoke of a new theory of gravity. He claimed to have measured the speed of cosmic rays from the star Antares and found them to travel fifty times faster than light. He attacked Einstein's theory of relativity with the confidence of a man who had not engaged seriously with the physics of the preceding three decades.

This is the part of the story that requires the most care. Some of Tesla's later claims were almost certainly wrong. His rejection of subatomic physics was wrong. His insistence that atoms were immutable was wrong. His assertion that mass could not be converted to energy was wrong — nuclear fission proved that decisively. He dismissed relativity with a confidence that his engagement with the underlying mathematics did not justify. He dismissed quantum mechanics for similar reasons. These were real errors, and they cost him credibility with the physics community at precisely the moment he most needed it.

His insistence on the ether is a more complicated matter. It is usually filed alongside the rest — further evidence that Tesla was trapped in a nineteenth-century framework. But Tesla's ether was not the rigid, static medium that Michelson and Morley had failed to detect in 1887. He described something altogether different: a dynamic, fluid-like medium filling all space, from which matter itself emerges as vortices of motion and to which it returns when the motion ceases. He called it the ether because that was the word available to him. What he was describing — a non-empty vacuum, a medium with fluid-like behavior, a substrate from which physical phenomena arise — sounds considerably less archaic when placed alongside the quantum vacuum of modern physics, which is not empty either, or alongside superfluid vacuum theory, which models space itself as a medium with fluid dynamics. Even Einstein, who supposedly killed the ether, conceded in his 1920 Leiden address that general relativity requires a medium "devoid of all mechanical and kinematical qualities, but which helps to determine mechanical and electromagnetic events." Tesla's word was outdated. Whether his intuition was is a question that modern physics has not settled as cleanly as most people assume.

The distinction matters for this story because it bears directly on how we evaluate Tesla's engineering. Every apparatus he ever built was designed on the assumption that energy propagates through a medium and that resonant coupling with that medium amplifies the effect. Every apparatus worked. If the foundational assumption were simply wrong — if space were truly empty and inert — there would be no mechanism for the resonant phenomena Tesla spent his career exploiting. The fact that his machines worked, consistently and at every scale, suggests that whatever he was tapping into, his functional understanding of it was correct, even if the theoretical vocabulary he used to describe it was not.

Not everything he said in these later years was wrong, and the temptation to dismiss all of it because some of it was mistaken should be resisted. His engineering intuitions remained acute. His understanding of electromagnetic phenomena, gained from decades of direct experimental work, was deeper than that of many theorists who dismissed him.
And some of his claims, particularly regarding directed energy and the transmission of power, described concepts that would later be pursued by military researchers with substantially larger budgets and substantially less imagination.

* * *

The most famous — and most controversial — of Tesla's later claims was the device he called teleforce. The press called it a death ray, a term Tesla detested. He insisted it was a death beam, and the distinction mattered to him: a ray diverges, losing intensity with distance; a beam remains focused. What Tesla described was a particle beam weapon.

The concept, as Tesla outlined it in a 1937 treatise titled "The New Art of Projecting Concentrated Non-dispersive Energy through the Natural Media," involved four components: an apparatus for producing rays in free air (eliminating the need for a vacuum), a mechanism for generating very great electrical force, a method for amplifying that force, and a new form of projecting apparatus. The principle was to accelerate tiny particles of tungsten or mercury to enormous velocities using electrostatic repulsion and project them in a concentrated stream through an open-ended vacuum tube with a gas jet seal at its exit. The particles would travel in a coherent beam, maintaining their density over distances Tesla claimed could reach two hundred miles.

He envisioned teleforce not as an offensive weapon but as a defensive one — a wall of destructive energy that could be erected along a nation's borders, rendering invasion impossible. In this, Tesla was consistent with the pacifist streak that had run through his thinking since the *Century* article of 1900, where he'd proposed a form of mutual deterrence as the path to permanent peace. The death beam was, in Tesla's conception, the weapon that would make all other weapons obsolete — not by winning wars but by making them unwinnable.

He offered the technology to multiple governments. The United States showed little interest. The Soviet Union reportedly paid Tesla twenty-five thousand dollars for preliminary plans, though the details of this transaction are disputed. Britain declined. No working prototype was ever built. The treatise describing the system's principles was not made public in Tesla's lifetime, but surfaced in 1984 in the archives of the Nikola Tesla Museum in Belgrade.

Was teleforce real? The honest answer is that the concept was technically coherent. Particle beam weapons have been researched extensively since Tesla's death, most notably during the Strategic Defense Initiative of the 1980s. The specific engineering Tesla described — accelerating charged particles to high velocity and projecting them through atmosphere over long distances — remains extremely difficult, and no one has achieved it at the scale Tesla envisioned. But the underlying physics is not fantasy. Whether Tesla himself had solved the engineering problems he claimed to have solved, with the limited resources available to him in a hotel room, is another question. The box he gave the Governor Clinton Hotel as collateral suggests that, at minimum, his claims about possessing a working model were not entirely straightforward.

* * *

In the autumn of 1937, at the age of eighty-one, Tesla left the Hotel New Yorker after midnight to make his regular walk to feed the pigeons. While crossing a street two blocks from the hotel, he was struck by a taxicab and thrown to the ground. His back was severely wrenched and three ribs were broken. Tesla refused to see a doctor, a habit of nearly a lifetime. He never fully recovered.

After the accident, he became increasingly confined to his hotel room. He still occasionally received visitors. He still made pronouncements to the press. But the walks grew rarer, and the world outside the thirty-third floor of the New Yorker receded further. He subsisted largely on milk, bread, honey, and vegetable juice. He asked the hotel staff who entered his room to remain at least three feet away from him. He placed a "do not disturb" sign on his door for days at a time.

* * *

On the night of January 7, 1943, at the age of eighty-six, Nikola Tesla died alone in Room 3327 of the Hotel New Yorker. His body was found the next day by a maid named Alice Monaghan, who entered the room after the "do not disturb" sign had been on the door for two days. The assistant medical examiner, H.W. Wembly, ruled the cause of death as coronary thrombosis. There were no suspicious circumstances.

The man whose polyphase system powered every city in the industrialized world, whose patents had made transatlantic communication possible, whose inventions had generated hundreds of millions of dollars in commercial value, died penniless and in debt, in a hotel room, alone.

* * *

What happened next says as much about Tesla's significance as anything he accomplished while alive.

Within two days of his death, the Federal Bureau of Investigation ordered the Office of Alien Property Custodian to seize Tesla's belongings. This was a remarkable action: Tesla had been an American citizen since 1891. But it was wartime, Tesla had spent years publicly claiming to possess a weapon of enormous destructive potential, and the government's concern was that his papers might fall into the hands of his nephew Sava Kosanović, a Yugoslav diplomat with suspected ties to the communist movement. Tesla's entire estate — papers, notebooks, equipment, personal effects — was gathered from the Hotel New Yorker and other storage locations and transported to the Manhattan Storage and Warehouse Company under government seal.

John G. Trump, a professor of electrical engineering at MIT and a technical consultant to the National Defense Research Committee, was assigned to evaluate the seized materials. Trump's assessment, delivered after a review of the papers, was blunt. Tesla's later work, he wrote, was "primarily of a speculative, philosophical, and somewhat promotional character," and did "not include new, sound, workable principles or methods for realizing such results." The particle beam weapon, Trump concluded, did not exist as a functioning technology.

But the military was not unanimous. Brigadier General L.C. Craigee, the first man to fly a jet aircraft for the U.S. military, reviewed the same material and reached a different conclusion. "There's something to this," he reportedly said. "The particle beam weapon is real." The government's position was thus split from the beginning: one faction dismissing Tesla's later work as the speculations of an elderly man, another taking it seriously enough to classify.

Tesla's estate was eventually released to Kosanović, who had been declared the rightful heir by a U.S. court in 1952. The papers and possessions were shipped to Belgrade in eighty trunks. Only sixty arrived. Whether the discrepancy represents efficient packing or government retention of selected materials has never been resolved. The papers now reside in the Nikola Tesla Museum in Belgrade. Tesla's ashes, cremated after a state funeral attended by two thousand people at the Cathedral of Saint John the Divine, are displayed there in a gold-plated sphere on a marble pedestal.

* * *

The funeral, on January 12, 1943, was a strangely grand affair for a man who had died destitute. Two thousand people filled the cathedral. New York City mayor Fiorello La Guardia read a eulogy over the radio, written by the Slovene-American author Louis Adamic, while a violinist played "Ave Maria." Nobel laureates and heads of state sent tributes. The Yugoslav and Czechoslovak governments both claimed him. Messages arrived from around the world.

The size of the funeral was the first indication that something had gone wrong with Tesla's reputation — that the obscurity of his final decades was not proportional to his significance. Here was a man whose patents, inventions, and engineering achievements had built the electrical infrastructure of the modern world, and he had died forgotten by the public and ignored by the institutions that ran on his work. The funeral attendance suggested that at least some portion of the scientific and engineering community understood this. They came not to bury an eccentric but to acknowledge a debt.

* * *

Less than six months after Tesla's death, the United States Supreme Court issued its ruling in *Marconi Wireless Telegraph Co. v. United States.* The Court invalidated the broad claims of Marconi's fundamental radio patent, number 763,772, finding that they had been anticipated by the earlier patents of Tesla, Oliver Lodge, and John Stone Stone. Tesla's patent 645,576, filed in September 1897, had described a four-circuit tuned wireless system before Marconi. The Court recognized this.

The engineering substance of the ruling deserves attention, because it goes to the heart of this book's argument. What the Supreme Court confirmed was that the foundational principle of radio — the use of resonant circuits tuned to a specific frequency at both transmitter and receiver — was Tesla's. Marconi's early apparatus was a crude spark-gap device with no tuning; it broadcast on all frequencies simultaneously, like shouting into a crowd. It was Tesla who recognized that if you tuned the transmitting circuit and the receiving circuit to the same resonant frequency, you could select one signal from the noise. This was not a minor refinement. It was the principle that made radio usable. Without it, only one station could transmit at a time. With it, thousands could operate simultaneously, each on its own frequency. Tuning — resonance — was the key, and the key was Tesla's.

Marconi's contribution was entrepreneurial, not scientific. He had the commercial instinct to build a business around wireless telegraphy, and he did it with energy and skill. But the apparatus he used was assembled from the published work of others — Tesla's resonant circuits, Lodge's coherer, Popov's antenna designs — and his central patent claim, when it finally reached the highest court in the land, was found to be anticipated by the very man whose priority he had spent decades denying. That Marconi received the Nobel Prize for Physics in 1909 for work built on Tesla's patents, while Tesla received nothing, is one of the enduring injustices of the history of science.

The timing was savage. Tesla had spent decades asserting his priority in wireless communication, and the legal system had finally confirmed it — nine months after he'd died in a hotel room without enough money to pay his bill. The ruling received almost no publicity. Its practical effect was merely to spare the U.S. government from paying damages to Marconi's company for wartime use of wireless technology. It did nothing to restore Tesla's reputation, and for the next several decades, most of the public continued to credit Marconi with the invention of radio. Many still do.

* * *

Tesla's work fell into relative obscurity after his death. His name appeared in textbooks as a unit of magnetic flux density — the tesla was adopted by the General Conference on Weights and Measures in 1960 — but the man himself was largely forgotten outside of engineering circles and a devoted following in the former Yugoslavia.

The rehabilitation began slowly. Biographies appeared. Engineers and physicists rediscovered his papers and recognized how much of the modern electrical world rested on his foundations. The internet, with its appetite for underdogs and its suspicion of established narratives, amplified Tesla's story into a cultural phenomenon. By the early twenty-first century, Tesla had become one of the most recognized inventors in history — more famous, in many ways, than he had been during the decades of his greatest productivity. In 2013, *Time* named him one of the hundred most significant figures of all time. Elon Musk named his electric car company after him.

The revival is welcome and largely deserved, but it carries its own distortions. The popular Tesla has become a mythological figure: the lone genius defeated by corporate greed, the mad scientist whose ideas were suppressed by powerful interests, the visionary who could have given the world free energy if only the villains had let him. This version is satisfying in the same way the "mad pigeon man" version is satisfying — it tells a clean story with clear heroes and villains. It is also, like that version, too simple.

* * *

Tesla was not a martyr. He was not a saint. He was not mad. He was an engineer of extraordinary gifts and extraordinary limitations, a man who understood electromagnetic resonance more deeply than anyone alive in his era and who failed, comprehensively, to manage money, relationships, or the politics of innovation. He was right about the rotating magnetic field. He was right about polyphase power. He was right about resonant amplification. He was right that the earth behaves as a resonant electromagnetic system. He was right that wireless communication and wireless power transmission were physically possible. He was wrong about relativity. He was wrong about the electron. He was wrong about the ease with which his World System could be funded, built, and operated. His insistence on the ether was, as we've seen, a more complicated matter — a nineteenth-century word wrapped around an intuition that modern physics has not entirely dismissed.

When Tesla talked about resonance, he was not speaking in metaphor. He was describing a set of physical principles he had spent forty years applying, testing, and confirming at every scale available to him. Those principles had produced the rotating magnetic field, the polyphase power system, the Tesla coil, and the magnifying transmitter. They had powered Niagara Falls, lit the Chicago World's Fair, and established the global electrical grid. Every apparatus Tesla designed on the basis of these principles had performed as calculated. This was not luck and it was not guesswork. It was engineering, conducted by a man who understood resonance with an intimacy that nobody around him shared.

And that is the key to understanding the last four decades of his life. Tesla was not descending into madness. He was operating with different principles. He saw electricity, resonance, and the behavior of energy in ways that his contemporaries could not follow — not because they were stupid, but because they had not done the work he had done. They had not stood inside the magnifying transmitter and felt the earth respond. They had not measured the planet's resonant frequency and found it matched their calculations. They had not transmitted power through the ground at distances that conventional theory said were impossible. Tesla had done all of these things. He was not speculating. He was reporting what he had observed. But the observations were so far outside the framework available to the people around him that they could not evaluate what he was telling them. They had no basis for saying he was right, and so they assumed he was wrong, and then they assumed he was delusional, and then they wrote stories about an old man and his pigeons.

The pigeons became the story because the real story was too large. A man claiming he could power the planet through resonant coupling with the earth's electromagnetic cavity is either a visionary of the highest order or a lunatic. There is no comfortable middle ground. And because the world had no means of evaluating the claim — because the only facility designed to test it had been demolished for scrap — it was easier, and kinder, to focus on the eccentricities. The birds. The hotel rooms. The napkins. The obsessive counting. These were things people could understand. The physics was not.

But Tesla's understanding did not diminish with age, even as his ability to demonstrate it did. He continued to think about resonance, about energy transmission, about the relationship between electrical systems and the planet that contained them. The birthday press conferences were not the ravings of a declining mind. They were the frustrated broadcasts of a man who had solved a problem that nobody else could see, and who had spent decades watching the world build an inferior version of the system he'd designed, using exactly the Hertzian radiation he'd told them was wasteful.

He watched radio develop along lines he considered fundamentally misguided. He watched power grids expand using copper wire to carry energy that he believed could travel through the earth itself. He watched the infrastructure of the twentieth century crystallize around principles he knew to be less efficient than the ones he had demonstrated and documented. And he could do nothing about it, because the demonstration that would have proven him right had been dynamited for seventeen hundred and fifty dollars in scrap metal.

* * *

The previous chapter made the case that Tesla's track record was flawless — that every apparatus he ever built had worked, and that Wardenclyffe was not an experiment but a machine designed to operate on principles already proven at smaller scale. That case stands. But what this chapter reveals is the human cost of what happens when a civilization cannot recognise what it is looking at.

Tesla spent his final years in a hotel room on the thirty-third floor, feeding pigeons and subsisting on a pension from the company that had made its fortune on his work. He was not broken. He was stranded. He had crossed into territory that his culture could not follow, and he had no way back and no means of pulling anyone forward. The world would catch up eventually — Schumann would confirm the earth's resonant cavity, the Supreme Court would confirm Tesla's radio patents, engineers would rediscover his papers and recognise how much of the modern world rested on his foundations — but by then Tesla would be ashes in a gold-plated sphere in a museum in Belgrade, and the tower would be rubble, and the question the tower was built to answer would still be open.

Not because the physics was wrong. Because no one else could see what Tesla could see. And in the absence of understanding, the world chose the story it could comprehend: a mad old man, alone in a hotel room, talking to birds.

It never occurred to anyone that the birds might not have been the point.

EPILOGUE

Whether you're reading this on a screen or on a printed page, it reaches you through an electrical world Nikola Tesla helped build. The generator that produced it, the transmission line that carried it, the transformer that stepped it down to a voltage your house can use — all of it runs on principles Tesla laid down before the twentieth century began. That's not mythology. It's the engineering record of a man whose every apparatus worked as designed. It is also, for most of the world, invisible. The system works so well that no one thinks about who designed it.

There is, however, something Tesla designed that you are not using. Nobody is. And unlike the system you take for granted every time you turn on a light, this one was never given the chance to work.

In 1899, at Colorado Springs, Tesla transmitted electrical power without wires using the ground as the transmission medium. He measured the earth's resonant frequency and found the planet responded like any other resonant system: drive it at the right frequency and it amplifies the signal. He documented everything in five hundred pages of laboratory notes, then designed a facility at Wardenclyffe to demonstrate the principle at planetary scale. The facility was demolished for scrap before the machine could be completed. That was 1917. In the century since, no one has replicated the experiment.

* * *

This book has tried to tell the story of a man, not a myth. The popular version — lone genius crushed by corporate villains — reduces a complex life to a fable. The opposite version — eccentric old man talking to pigeons

— uses the surface of his later years to avoid engaging with the substance of his work. Neither version is honest.

The substance is this. Tesla applied principles of electrical resonance across four decades, at every scale from a laboratory bench to the North American power grid. He was never wrong. Every theory produced a working machine. Every machine performed as calculated. And then he said he could do it with the planet.

And he began to. And the money stopped. And the tower came down. And a hundred years passed.

* * *

We built the modern world on the half of Tesla's vision that was easy to monetize. Alternating current can be generated, transmitted through copper wire, metered at the point of delivery, and billed. It requires vast infrastructure, and vast infrastructure requires capital, and capital requires a return. The system works.

The other half — energy delivered through the body of the planet to anyone with a tuned receiver — was never tested. Not because the physics was refuted. Because the business model was unclear. Because the man who designed it couldn't pay his hotel bill. Because the tower was worth more as scrap than as science.

The only reason we do not know whether the earth can serve as a medium for wireless power transmission is that seventeen hundred and fifty dollars in scrap metal was more persuasive than the track record of the most consistently correct engineer of the modern age.

* * *

Consider what a century of development might have produced.

If Tesla's wireless power system had worked at planetary scale — if a network of transmitting stations had been built in the early twentieth century, each coupling energy into the earth and making it available to any tuned receiver anywhere on the surface — the engineering consequences cascade in directions that are difficult to overstate.

Start with the most obvious. The battery is the bottleneck of the modern world. Every portable technology we possess — every phone, every laptop, every electric vehicle, every drone, every medical implant, every satellite — is designed around a single constraint: it must carry its own energy supply. The entire history of portable electronics is a history of managing that constraint. Make the battery smaller. Make it lighter. Make it last longer. Make it charge faster. Billions of dollars and decades of engineering have been poured into a problem that, in a world with ambient wireless power, does not exist.

The electric car is the clearest example. Tesla Motors — the company named after the man, not the man himself — spent twenty years and tens of billions of dollars solving the problem of how to store enough energy in a vehicle to make it commercially useful. The battery pack in a modern electric car accounts for a third of its weight, a third of its cost, and nearly all of its range anxiety. It degrades with use. It requires rare earth minerals extracted at enormous environmental cost. It takes time to charge. Remove the battery and the electric car becomes lighter, cheaper, simpler, and capable of unlimited range. It draws power from the ground beneath it the way a tram draws power from an overhead wire, except there is no wire and no track. It just drives.

Now extend the principle. An electric aircraft without batteries has no weight-to-energy problem. The reason electric aviation remains marginal in the twenty-first century is that batteries are too heavy to make long-range flight practical. Remove the constraint and electric flight becomes feasible — not in 2025, but in 1925. Drones with unlimited endurance. Cargo delivery without fuel costs. Air travel without kerosene.

A phone without a battery never dies. It draws power from the ambient field the way a crystal radio draws signal from the air. The entire miniaturization race of the past fifty years — driven largely by the need to fit more capability around a smaller battery — takes a completely different path. So does the economics of personal technology: if the device doesn't need to be replaced when the battery degrades, the cycle of planned obsolescence that drives the consumer electronics industry loses its primary mechanism.

The power grid itself — the vast physical network of transmission towers, copper cable, substations, transformers, and metering equipment that stretches across every developed landscape on earth — doesn't exist. It isn't needed. There are no power lines. There are no utility poles. Rural electrification, which took decades and still hasn't reached hundreds of millions of people, happens instantly. A village in central Africa receives power at the same moment as Manhattan, because there is nothing to build between the source and the receiver. The staggering inequality of energy access that defines the modern world — in which a billion people still have no reliable electricity — is an artifact of a distribution system that requires physical infrastructure between every generator and every user. Remove the infrastructure requirement and the inequality collapses.

And then there is the thing that no one talks about when they talk about Tesla, because the implications are too large.

If energy is not delivered through combustion, the petroleum economy never becomes the organizing principle of the twentieth century.

The internal combustion engine, which dominated transportation for a hundred years not because it was the best technology but because liquid fuel was the most portable energy source available, never achieves that dominance. The geopolitical architecture built on controlling oil supply routes — the wars, the alliances, the petrodollar, the strategic significance of the Middle East — forms differently, or doesn't form at all. The carbon emissions that followed from a century of burning fossil fuels for energy that could, in principle, have been delivered without burning anything follow a different trajectory. The climate crisis, as we know it, may not exist in the same form.

This is not utopianism. Wireless power transmission would have produced its own problems, its own inequalities, its own geopolitical struggles over who controls the transmitters and who sets the frequencies. The point is not that Tesla's world would have been perfect. The point is that it would have been — in ways that touch transportation, communication, energy access, international relations, and the composition of the atmosphere — fundamentally different from the one we built. And we built the one we built not because we tested both options and chose the better one, but because we tested one option and demolished the other for scrap.

* * *

This is not a conspiracy theory about suppressed technology. It is something simpler and more uncomfortable. A man with a perfect engineering record told us something specific about how the planet works, and we have spent a century not checking.

The tools exist now that didn't exist in Tesla's time. The materials science has caught up. The computational power to model earth-ionosphere cavity resonance is available on a desktop. The cost of a serious experimental programme would be trivial compared to the annual budgets of the energy companies that run on his other inventions. It would not require a tower. It would not require a Morgan. It would require only the willingness to take seriously a man who was never wrong, and to find out whether he was wrong this one time, about this one thing.

If he was wrong, we lose nothing but a persistent question. If he was right, we have wasted a century.

* * *

Tesla saw the rotating magnetic field complete in a park in Budapest. He drew it in the dirt with a stick. Everything that followed — the motor, the grid, the coil, the magnifying transmitter, the tower — was the same vision at increasing scale. He saw one more thing: the planet itself as a resonant body, capable of carrying energy to every point on its surface. He spent forty years insisting it would work. He died in a hotel room before he could show us.

The question is not whether we believe him. The question is whether we are willing to look.

*"The present is theirs; the future,
for which I really worked, is mine."*

— *Nikola Tesla, 1927*

Made in the USA
Middletown, DE
05 November 2026

36541633R00102